SpringerBriefs in Computer Science

Series Editors

Stan Zdonik
Peng Ning
Shashi Shekhar
Jonathan Katz
Xindong Wu
Lakhmi C. Jain
David Padua
Xuemin Shen
Borko Furht
V. S. Subrahmanian
Martial Hebert
Katsushi Ikeuchi
Bruno Siciliano

For further volumes:
http://www.springer.com/series/10028

Miles Hansard · Seungkyu Lee
Ouk Choi · Radu Horaud

Time-of-Flight Cameras

Principles, Methods and Applications

 Springer

Miles Hansard
Electronic Engineering
  and Computer Science
Queen Mary, University of London
London
UK

Seungkyu Lee
Samsung Advanced Institute
  of Technology
Yongin-si
Kyonggi-do
Republic of South Korea

Ouk Choi
Samsung Advanced Institute
  of Technology
Yongin-si
Kyonggi-do
Republic of South Korea

Radu Horaud
INRIA Grenoble Rhône-Alpes
Montbonnot Saint-Martin
France

ISSN 2191-5768   ISSN 2191-5776   (electronic)
ISBN 978-1-4471-4657-5   ISBN 978-1-4471-4658-2   (eBook)
DOI 10.1007/978-1-4471-4658-2
Springer London Heidelberg New York Dordrecht

Library of Congress Control Number: 2012950373

© Miles Hansard 2013
This work is subject to copyright. All rights are reserved by the Publisher, whether the whole or part of the material is concerned, specifically the rights of translation, reprinting, reuse of illustrations, recitation, broadcasting, reproduction on microfilms or in any other physical way, and transmission or information storage and retrieval, electronic adaptation, computer software, or by similar or dissimilar methodology now known or hereafter developed. Exempted from this legal reservation are brief excerpts in connection with reviews or scholarly analysis or material supplied specifically for the purpose of being entered and executed on a computer system, for exclusive use by the purchaser of the work. Duplication of this publication or parts thereof is permitted only under the provisions of the Copyright Law of the Publisher's location, in its current version, and permission for use must always be obtained from Springer. Permissions for use may be obtained through RightsLink at the Copyright Clearance Center. Violations are liable to prosecution under the respective Copyright Law.
The use of general descriptive names, registered names, trademarks, service marks, etc. in this publication does not imply, even in the absence of a specific statement, that such names are exempt from the relevant protective laws and regulations and therefore free for general use.
While the advice and information in this book are believed to be true and accurate at the date of publication, neither the authors nor the editors nor the publisher can accept any legal responsibility for any errors or omissions that may be made. The publisher makes no warranty, express or implied, with respect to the material contained herein.

Printed on acid-free paper

Springer is part of Springer Science+Business Media (www.springer.com)

Preface

This book describes a variety of recent research into time-of-flight imaging. Time-of-flight cameras are used to estimate 3D scene structure directly, in a way that complements traditional multiple-view reconstruction methods. The first two chapters of the book explain the underlying measurement principle, and examine the associated sources of error and ambiguity. Chapters 3 and 4 are concerned with the geometric calibration of time-of-flight cameras, particularly when used in combination with ordinary color cameras. The final chapter shows how to use time-of-flight data in conjunction with traditional stereo matching techniques. The five chapters, together, describe a complete depth and color 3D reconstruction pipeline. This book will be useful to new researchers in the field of depth imaging, as well as to those who are working on systems that combine color and time-of-flight cameras.

Acknowledgments

The work presented in this book has been partially supported by a co-operative research project between the 3D Mixed Reality Group at the Samsung Advanced Institute of Technology in Seoul, South Korea and the Perception group at INRIA Grenoble Rhône-Alpes in Montbonnot Saint-Martin, France.

The authors would like to thank Michel Amat for his contributions to Chaps. 3 and 4, as well as Jan Cech and Vineet Gandhi for their contributions to Chap. 5.

Contents

1 Characterization of Time-of-Flight Data 1
   1.1 Introduction 1
   1.2 Principles of Depth Measurement 2
   1.3 Depth-Image Enhancement. 3
       1.3.1 Systematic Depth Error 4
       1.3.2 Nonsystematic Depth Error 5
       1.3.3 Motion Blur. 5
   1.4 Evaluation of Time-of-Flight and Structured-Light Data. 12
       1.4.1 Depth Sensors. 13
       1.4.2 Standard Depth Data Set 14
       1.4.3 Experiments and Analysis 18
       1.4.4 Enhancement 22
   1.5 Conclusions 25
   References 26

2 Disambiguation of Time-of-Flight Data 29
   2.1 Introduction 29
   2.2 Phase Unwrapping from a Single Depth Map 30
       2.2.1 Deterministic Methods. 35
       2.2.2 Probabilistic Methods 36
       2.2.3 Discussion 38
   2.3 Phase Unwrapping from Multiple Depth Maps 38
       2.3.1 Single-Camera Methods. 39
       2.3.2 Multicamera Methods 40
       2.3.3 Discussion 42
   2.4 Conclusions 42
   References 43

3 Calibration of Time-of-Flight Cameras ... 45
3.1 Introduction ... 45
3.2 Camera Model ... 46
3.3 Board Detection ... 46
    3.3.1 Overview ... 48
    3.3.2 Preprocessing ... 49
    3.3.3 Gradient Clustering ... 49
    3.3.4 Local Coordinates ... 51
    3.3.5 Hough Transform ... 51
    3.3.6 Hough Analysis ... 53
    3.3.7 Example Results ... 55
3.4 Conclusions ... 56
References ... 58

4 Alignment of Time-of-Flight and Stereoscopic Data ... 59
4.1 Introduction ... 59
4.2 Methods ... 62
    4.2.1 Projective Reconstruction ... 63
    4.2.2 Range Fitting ... 63
    4.2.3 Point-Based Alignment ... 64
    4.2.4 Plane-Based Alignment ... 66
    4.2.5 Multisystem Alignment ... 68
4.3 Evaluation ... 69
    4.3.1 Calibration Error ... 70
    4.3.2 Total Error ... 70
4.4 Conclusions ... 72
References ... 74

5 A Mixed Time-of-Flight and Stereoscopic Camera System ... 77
5.1 Introduction ... 77
    5.1.1 Related Work ... 78
    5.1.2 Chapter Contributions ... 81
5.2 The Proposed ToF-Stereo Algorithm ... 82
    5.2.1 The Growing Procedure ... 82
    5.2.2 ToF Seeds and Their Refinement ... 83
    5.2.3 Similarity Statistic Based on Sensor Fusion ... 86
5.3 Experiments ... 88
    5.3.1 Real-Data Experiments ... 88
    5.3.2 Comparison Between ToF Map and Estimated Disparity Map ... 90
    5.3.3 Ground-Truth Evaluation ... 91
    5.3.4 Computational Costs ... 92
5.4 Conclusions ... 94
References ... 94

Chapter 1
Characterization of Time-of-Flight Data

Abstract This chapter introduces the principles and difficulties of time-of-flight depth measurement. The depth images that are produced by time-of-flight cameras suffer from characteristic problems, which are divided into the following two classes. First, there are systematic errors, such as noise and ambiguity, which are directly related to the sensor. Second, there are nonsystematic errors, such as scattering and motion blur, which are more strongly related to the scene content. It is shown that these errors are often quite different from those observed in ordinary color images. The case of motion blur, which is particularly problematic, is examined in detail. A practical methodology for investigating the performance of depth cameras is presented. Time-of-flight devices are compared to structured-light systems, and the problems posed by specular and translucent materials are investigated.

Keywords Depth-cameras · Time-of-Flight principle · Motion blur · Depth errors

1.1 Introduction

Time-of-Flight (ToF) cameras produce a *depth image*, each pixel of which encodes the distance to the corresponding point in the scene. These cameras can be used to estimate 3D structure directly, without the help of traditional computer-vision algorithms. There are many practical applications for this new sensing modality, including robot navigation [31, 37, 50], 3D reconstruction [17], and human–machine interaction [9, 45]. ToF cameras work by measuring the phase delay of reflected infrared (IR) light. This is not the only way to estimate depth; for example, an IR *structured-light* pattern can be projected onto the scene, in order to facilitate visual triangulation [44]. Devices of this type, such as the Kinect [12], share many applications with ToF cameras [8, 33, 34, 36, 43].

The unique sensing architecture of the TOF camera means that a raw depth image contains both systematic and nonsystematic bias that has to be resolved for robust depth imaging [11]. Specifically, there are problems of low depth precision and low spatial resolution, as well as errors caused by radiometric, geometric, and illumination variations. For example, measurement accuracy is limited by the power of the emitted IR signal, which is usually rather low compared to daylight, such that the latter contaminates the reflected signal. The amplitude of the reflected IR also varies according to the material and color of the object surface.

Another critical problem with TOF depth images is *motion blur*, caused by either camera or object motion. The motion blur of TOF data shows unique characteristics, compared to that of conventional color cameras. Both the depth accuracy and the frame rate are limited by the required integration time of the depth camera. Longer integration time usually allows higher accuracy of depth measurement. For static objects, we may therefore want to decrease the frame rate in order to obtain higher measurement accuracies from longer integration times. On the other hand, capturing a moving object at fixed frame rate imposes a limit on the integration time.

In this chapter, we discuss depth-image noise and error sources, and perform a comparative analysis of TOF and structured-light systems. First, the TOF depth-measurement principle will be reviewed.

1.2 Principles of Depth Measurement

Figure 1.1 illustrates the principle of TOF depth sensing. An IR wave indicated in red is directed to the target object, and the sensor detects the reflected IR component. By measuring the phase difference between the radiated and reflected IR waves, we can calculate the distance to the object. The phase difference is calculated from the relation between four different electric charge values as shown in Fig. 1.2. The four phase control signals have 90 degree phase delays from each other. They determine the collection of electrons from the accepted IR. The four resulting electric charge values are used to estimate the phase difference t_d as

$$t_d = \arctan\left(\frac{Q_3 - Q_4}{Q_1 - Q_2}\right) \quad (1.1)$$

where Q_1 to Q_4 represent the amount of electric charge for the control signals C_1 to C_4, respectively [11, 20, 23]. The corresponding distance d can then be calculated, using c the speed of light and f the signal frequency:

$$d = \frac{c}{2f}\frac{t_d}{2\pi}. \quad (1.2)$$

1.2 Principles of Depth Measurement

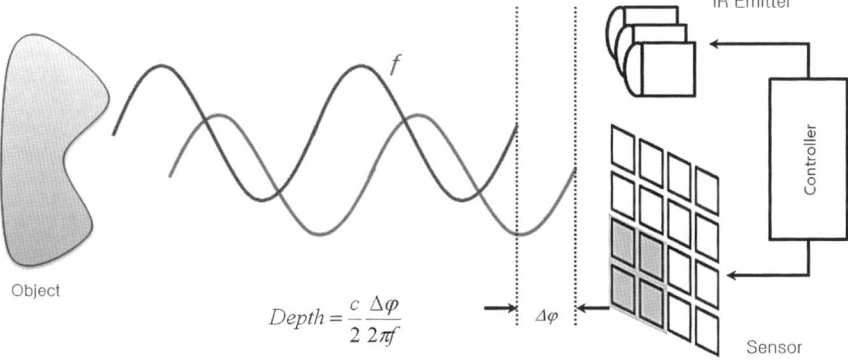

Fig. 1.1 The principle of TOF depth camera [11, 20, 23]: the phase delay between emitted and reflected IR signals are measured to calculate the distance from each sensor pixel to target objects

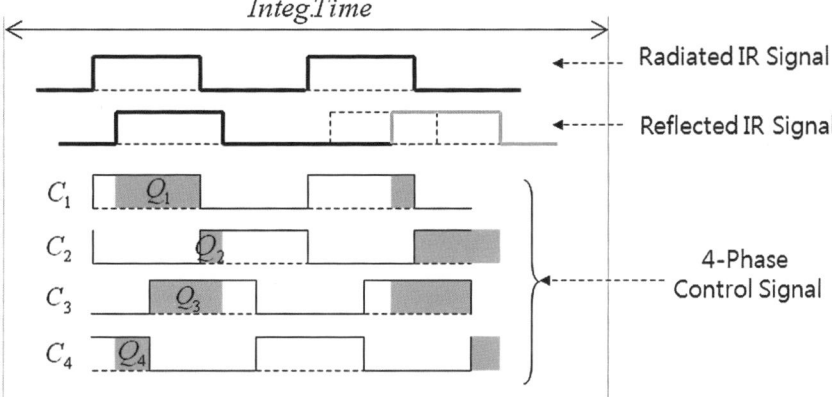

Fig. 1.2 Depth can be calculated by measuring the phase delay between radiated and reflected IR signals. The quantities Q_1 to Q_4 represent the amount of electric charge for control signals C_1 to C_4 respectively

Here, the quantity $c/(2f)$ is the maximum distance that can be measured without ambiguity, as will be explained in Chap. 2.

1.3 Depth-Image Enhancement

This section describes the characteristic sources of error in TOF imaging. Some methods for reducing these errors are discussed. The case of motion blur, which is particularly problematic, is considered in detail.

Fig. 1.3 Systematic noise and error: these errors come from the ToF principle of depth measurement. **a** Integration time error: longer integration time shows higher depth accuracy (*right*) than shorter integration time (*left*). **b** IR amplitude error: 3D points of the same depth (chessboard on the *left*) show different IR amplitudes (chessboard on the *right*) according to the color of the target object

1.3.1 Systematic Depth Error

From the principle and architecture of ToF sensing, depth cameras suffer from several systematic errors such as IR demodulation error, integration time error, amplitude ambiguity, and temperature error [11]. As shown in Fig. 1.3a, longer integration increases signal-to-noise ratio, which, however, is also related to the frame rate. Figure 1.3b shows that the amplitude of the reflected IR signal varies according to the color of the target object as well as the distance from the camera. The ambiguity of IR amplitude introduces noise into the depth calculation.

1.3.2 Nonsystematic Depth Error

Light scattering [32] gives rise to artifacts in the depth image, due to the low sensitivity of the device. As shown in Fig. 1.4a, close objects (causing IR saturation) in the lower right part of the depth image introduce depth distortion in other regions, as indicated by dashed circles. Multipath error [13] occurs when a depth calculation in a sensor pixel is an superposition of multiple reflected IR signals. This effect becomes serious around the concave corner region as shown in Fig. 1.4b. Object boundary ambiguity [35] becomes serious when we want to reconstruct a 3D scene based on the depth image. Depth pixels near boundaries fall in between foreground and background, giving rise to 3D structure distortion.

1.3.3 Motion Blur

Motion blur, caused by camera or target object motions, is a critical error source for online 3D capturing and reconstruction with ToF cameras. Because the 3D depth measurement is used to reconstruct the 3D geometry of scene, blurred regions in a depth image lead to serious distortions in the subsequent 3D reconstruction. In this section, we study the theory of ToF depth sensors and analyze how motion blur occurs, and what it looks like. Due the its unique sensing architecture, motion blur in the ToF depth camera is quite different from that of color cameras, which means that existing deblurring methods are inapplicable.

The motion blur observed in a depth image has a different appearance from that in a color image. Color motion blur shows smooth color transitions between foreground and background regions [46, 47, 51]. On the other hand, depth motion blur tends to present overshoot or undershoot in depth-transition regions. This is due to the different sensing architecture in ToF cameras, as opposed to conventional color cameras. The ToF depth camera emits an IR signal of a specific frequency, and measures the phase difference between the emitted and reflected IR signals to obtain the depth from the camera to objects. While calculating the depth value from the IR measurements, we need to perform a nonlinear transformation. Due to this architectural difference, the smooth error in phase measurement can cause uneven error terms, such as overshoot or undershoot. As a result, such an architectural difference between depth and color cameras makes the previous color image deblurring algorithms inapplicable to depth images.

Special cases of this problem have been studied elsewhere. Hussmann et al. [19] introduce a motion blur detection technique on a conveyor belt, in the presence of a single directional motion. Lottner et al. [28] propose an internal sensor control signal based blur detection method that is inappropriate in general settings. Lindner et al. [26] model the ToF motion blur in the depth image, to compensate for the artifact. However, they introduce a simple blur case without considering the ToF

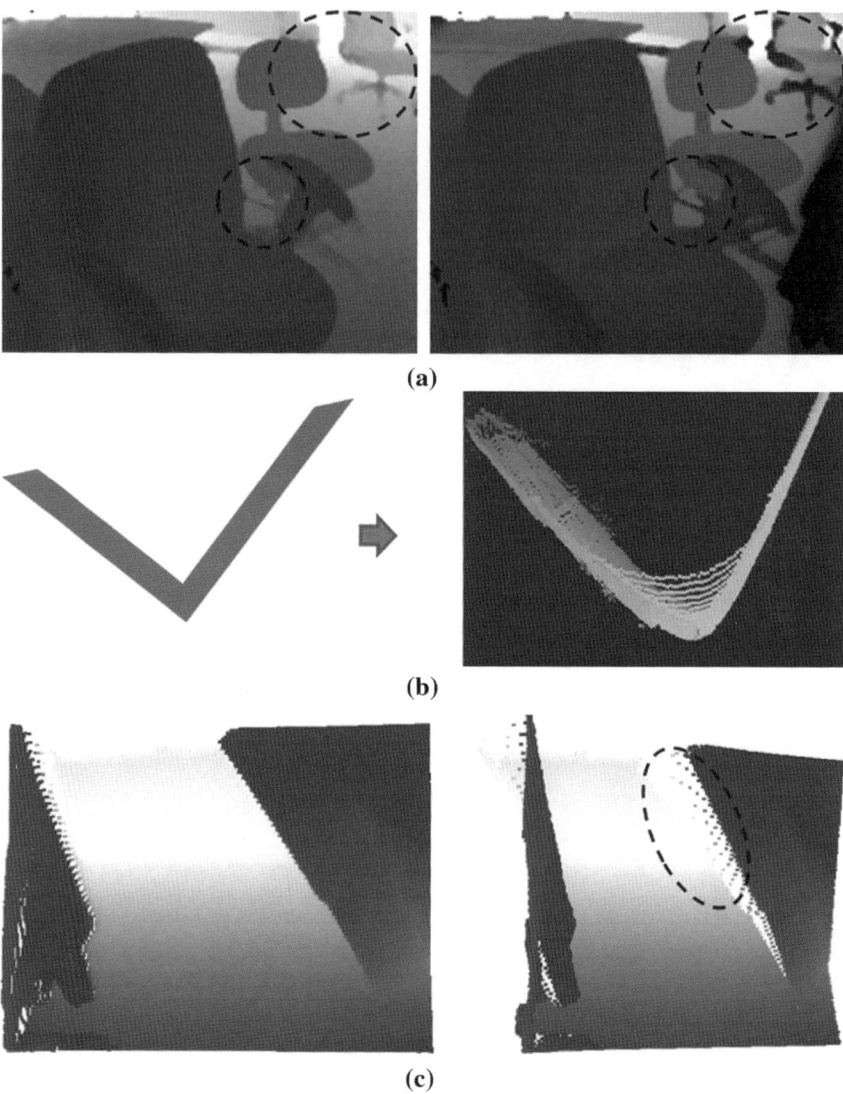

Fig. 1.4 Nonsystematic noise and error: based on the depth-sensing principle, scene structure may cause characteristic errors. **a** Light scattering: IR saturation in the lower right part of the depth image causes depth distortion in other parts, as indicated by *dashed circles*. **b** Multipath error: the region inside the concave corner is affected, and shows distorted depth measurements. **c** Object boundary ambiguity: several depth points on an object boundary are located in between foreground and background, resulting in 3D structure distortion

1.3 Depth-Image Enhancement

Fig. 1.5 ToF depth motion-blur due to movement of the target object

principle of depth sensing. Lee et al. [24, 25] examine the principle of ToF depth blur artifacts, and propose systematic blur detection and deblurring methods.

Based on the depth-sensing principle, we will investigate how motion blur occurs, and what are its characteristics. Let us assume that any motion from camera or object occurs during the integration time, which changes the phase difference of the reflected IR as indicated by the gray color in Fig. 1.2. In order to collect enough electric charge Q_1 to Q_4 to calculate depth (1.1), we have to maintain a sufficient integration time. According to the architecture type, integration time can vary, but the integration time is the major portion of the processing time. Suppose that n cycles are used for the depth calculation. In general, we repeat the calculation n times during the integration time to increase the signal-to-noise ratio, and so

$$t_d = \arctan\left(\frac{nQ_3 - nQ_4}{nQ_1 - nQ_2}\right) \quad (1.3)$$

where Q_1 to Q_4 represent the amount of electric charge for the control signals C_1 to C_4, respectively (cf. Eq. 1.1 and Fig. 1.2). The depth calculation formulation (1.3) expects that the reflected IR during the integration time comes from a single 3D point of the scene. However, if there is any camera or object motion during the integration time, the calculated depth will be corrupted. Figure 1.5 shows an example of this situation. The red dot represents a sensor pixel of the same location. Due the motion of the chair, the red dot sees both foreground and background sequentially within its integration time, causing a false depth calculation as shown in the third image in Fig. 1.5. The spatial collection of these false-depth points looks like blur around moving object boundaries, where significant depth changes are present.

Figure 1.6 illustrates what occurs at motion blur pixels in the '2-tab' architecture, where only two electric charge values are available. In other words, only $Q_1 - Q_2$ and $Q_3 - Q_4$ values are stored, instead of all separate Q values. Figure 1.6a is the case where no motion blur occurs. In the plot of $Q_1 - Q_2$ versus $Q_3 - Q_4$ in the third column, all possible regular depth values are indicated by blue points, making a diamond shape. If there is a point deviating from it, as an example shown in Fig. 1.6b, it means that their is a problem in between the charge values Q_1 to Q_4. As we already explained in Fig. 1.2, this happens when there exist multiple reflected signals with different phase values. Let us assume that a new reflected signal, of a different phase value, comes in from the mth cycle out of a total of n cycles during the first half *or* second half of the integration time. A new depth is then obtained as

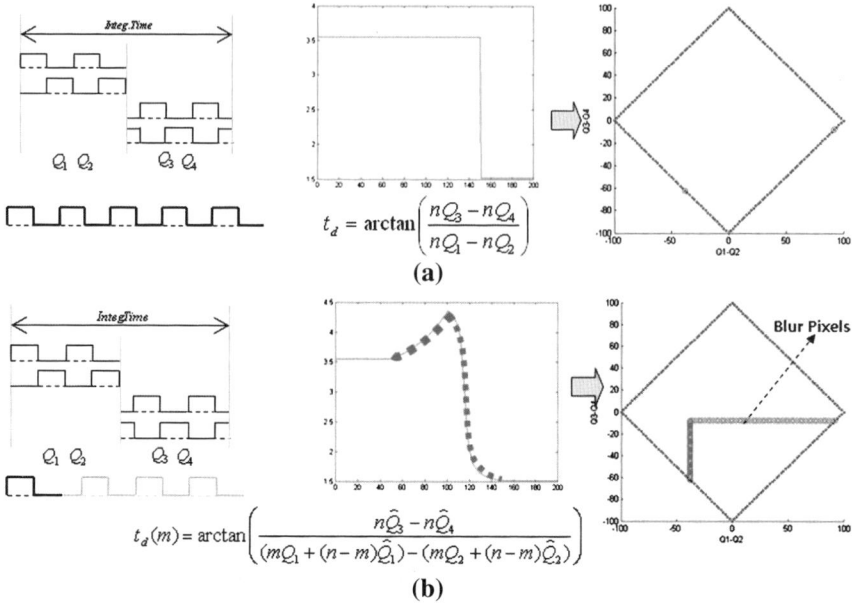

Fig. 1.6 ToF depth-sensing and temporal integration

$$t_d(m) = \arctan\left(\frac{n\hat{Q}_3 - n\hat{Q}_4}{(mQ_1 + (n-m)\hat{Q}_1) - (mQ_2 + (n-m)\hat{Q}_2)}\right) \quad (1.4)$$

$$t_d(m) = \arctan\left(\frac{(mQ_3 + (n-m)\hat{Q}_3) - (mQ_4 + (n-m)\hat{Q}_4)}{n\hat{Q}_1 - n\hat{Q}_2}\right) \quad (1.5)$$

in the first or second half of the integration time, respectively. Using the depth calculation formulation Eq. (1.1), we simulate all possible blur models. Figure 1.7 illustrates several examples of depth images taken by ToF cameras, having depth value transitions in motion blur regions. Actual depth values along the blue and red cuts in each image are presented in the following plots. The motion blurs of depth images in the middle show unusual peaks (blue cut) which cannot be observed in conventional color motion blur. Figure 1.8 shows how motion blur appears in 2-tap case. In the second phase where control signals C_3 and C_4 collect electric charges, the reflected IR signal is a mixture of background and foreground. Unlike color motion blurs, depth motion blurs often show overshoot or undershoot in their transition between foreground and background regions. This means that motion blurs result in higher or lower calculated depth than all near foreground and background depth values, as demonstrated in Fig. 1.9.

In order to verify this characteristic situation, we further investigate the depth calculation formulation in Eq. 1.5. First, we re-express Eq. 1.4 as

1.3 Depth-Image Enhancement

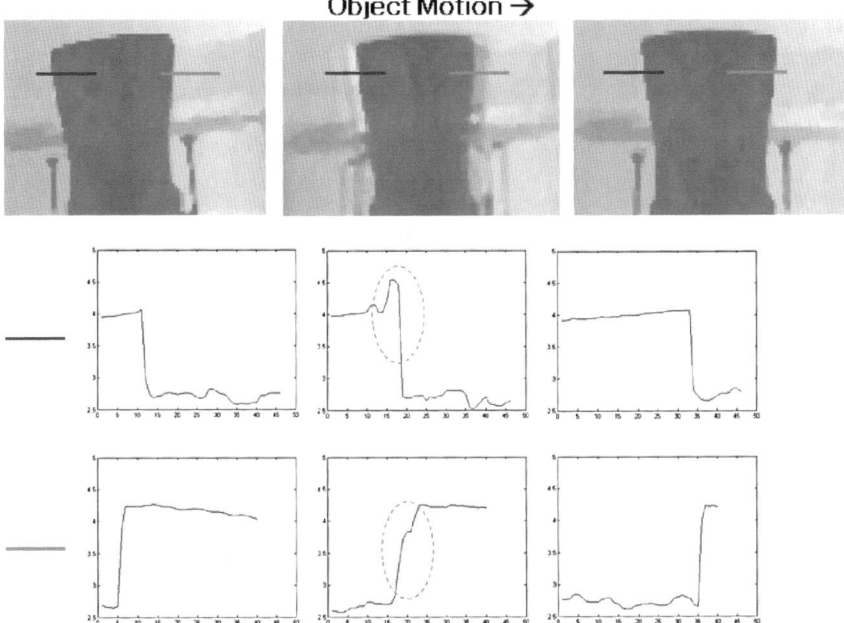

Fig. 1.7 Sample depth value transitions from depth motion blur images captured by an SR4000 ToF camera

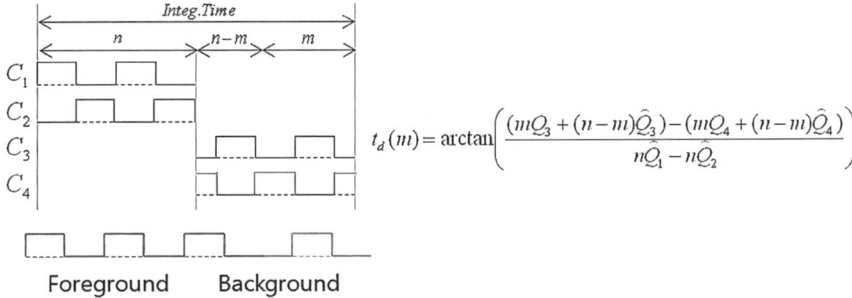

Fig. 1.8 Depth motion blur in 2-tap case

$$t_d(m) = \arctan\left(\frac{n\hat{Q}_3 - n\hat{Q}_4}{m(Q_1 - \hat{Q}_1 - Q_2 + \hat{Q}_2) + n(\hat{Q}_1 - \hat{Q}_2)}\right) \quad (1.6)$$

The first derivative of the Eq. 1.6 is zero, meaning local maxima or local minima, under the following conditions:

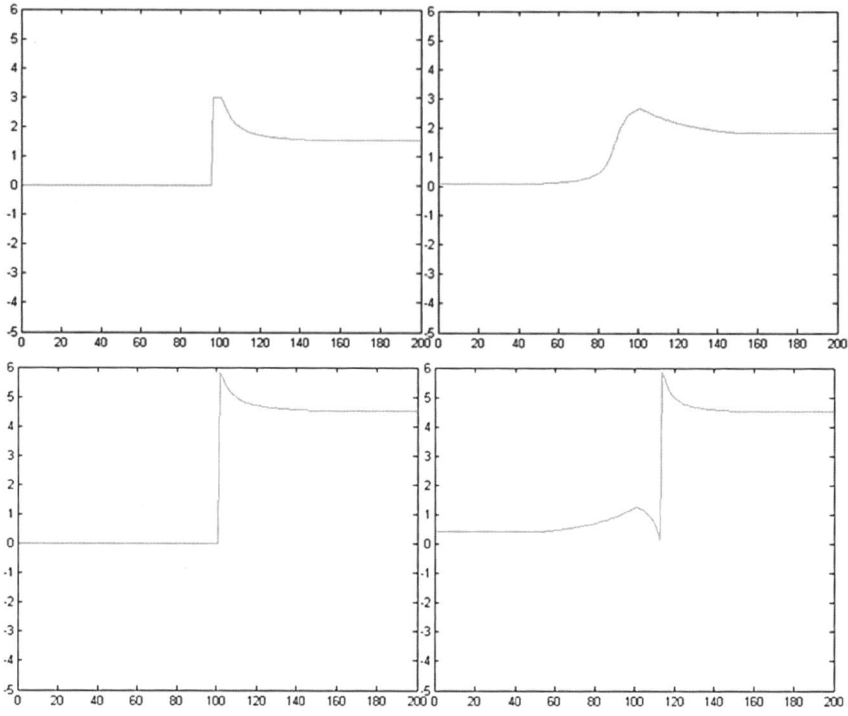

Fig. 1.9 ToF depth motion blur simulation results

$$t'_d(m) = \cfrac{1}{1+\left(\cfrac{n\hat{Q}_3-n\hat{Q}_4}{m(Q_1-\hat{Q}_1-Q_2+\hat{Q}_2)+n(\hat{Q}_1-\hat{Q}_2)}\right)^2} \qquad (1.7)$$

$$= \frac{(m(Q_1-\hat{Q}_1-Q_2+\hat{Q}_2)+n(\hat{Q}_1-\hat{Q}_2))^2}{(n\hat{Q}_3-n\hat{Q}_4)^2+(m(Q_1-\hat{Q}_1-Q_2+\hat{Q}_2)+n(\hat{Q}_1-\hat{Q}_2))^2} = 0$$

$$m = n\frac{\hat{Q}_2-\hat{Q}_1}{Q_1-\hat{Q}_1-Q_2+\hat{Q}_2} = n\frac{1-2\hat{Q}_1}{2Q_1-2\hat{Q}_1} \qquad (1.8)$$

Figure 1.10 shows that statistically half of all cases have overshoots or undershoots. In a similar manner, the motion blur model of 1-tap (Eq. 1.9) and 4-tap (Eq. 1.10) cases can be derived. Because a single memory is assigned for recording the electric charge value of four control signals, the 1-tap case has four different formulations upon each phase transition:

1.3 Depth-Image Enhancement

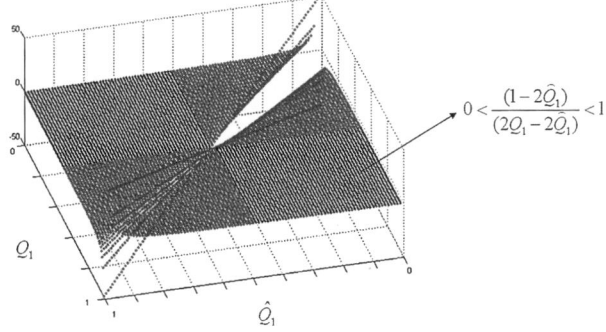

Fig. 1.10 Half of the all motion blur cases make local peaks

$$t_d(m) = \arctan\left(\frac{n\hat{Q}_3 - n\hat{Q}_4}{(mQ_1 + (n-m)\hat{Q}_1) - n\hat{Q}_2}\right)$$

$$t_d(m) = \arctan\left(\frac{n\hat{Q}_3 - n\hat{Q}_4}{nQ_1 - (mQ_2 + (n-m)\hat{Q}_2)}\right)$$

$$t_d(m) = \arctan\left(\frac{(mQ_3 + (n-m)\hat{Q}_3) - n\hat{Q}_4}{n\hat{Q}_1 - n\hat{Q}_2}\right)$$

$$t_d(m) = \arctan\left(\frac{nQ_3 - (mQ_4 + (n-m)\hat{Q}_4)}{n\hat{Q}_1 - n\hat{Q}_2}\right) \quad (1.9)$$

On the other hand, the 4-tap case only requires a single formulation, which is:

$$t_d(m) = \arctan\left(\frac{(mQ_3 + (n-m)\hat{Q}_3) - (mQ_4 + (n-m)\hat{Q}_4)}{(mQ_1 + (n-m)\hat{Q}_1) - (mQ_2 + (n-m)\hat{Q}_2)}\right) \quad (1.10)$$

Now, by investigating the relation between control signals, any corrupted depth easily can be identified. From the relation between Q_1 and Q_4, we find the following relation:

$$Q_1 + Q_2 = Q_3 + Q_4 = K. \quad (1.11)$$

Let us call this the *Plus Rule*, where K is the total amount of charged electrons. Another relation is the following formulation, called the *Minus Rule*:

$$|Q_1 - Q_2| + |Q_3 - Q_4| = K. \quad (1.12)$$

In fact, neither formulation exclusively represents motion blur. Any other event that can break the relation between the control signals, and can be detected by one of the rules, is an error which must be detected and corrected. We conclude that ToF motion blur can be detected by one or more of these rules.

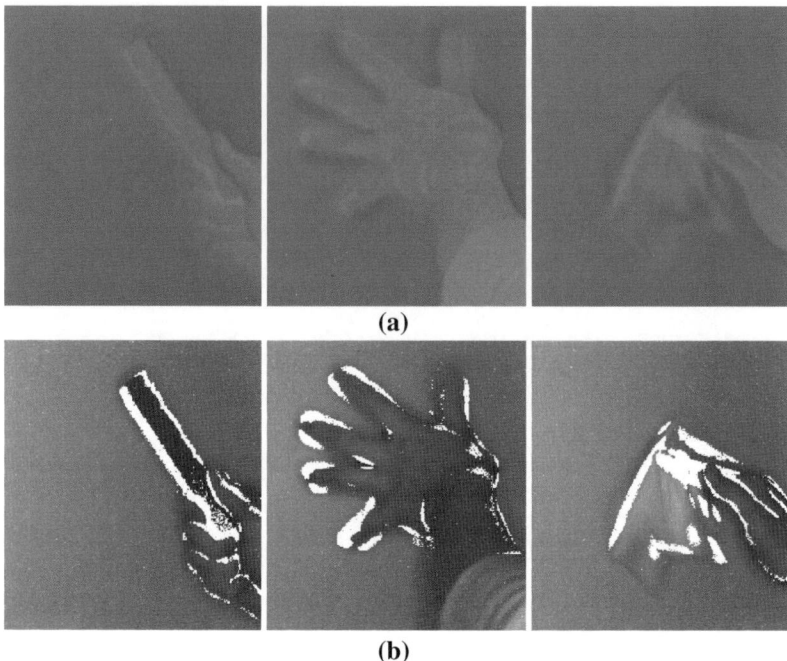

Fig. 1.11 Depth-image motion blur detection results by the proposed method. **a** Depth images with motion blur. **b** Intensity images with detected motion blur regions (indicated by *white color*)

Figure 1.11a shows depth-image samples with motion blur artifacts due to various object motions such as rigid body, multiple body, and deforming body motions, respectively. Motion blur occurs not just around object boundaries; inside an object, any depth differences that are observed within the integration time will also cause motion blur. Figure 1.11b shows detected motion blur regions indicated by white color on respective depth and intensity images, by the method proposed in [24]. This is very straightforward but effective and fast method, which is fit for hardware implementation without any additional frame memory or processing time.

1.4 Evaluation of Time-of-Flight and Structured-Light Data

The enhancement of ToF and structured-light (e.g., Kinect [44]) data is an important topic, owing to the physical limitations of these devices (as described in Sect. 1.3). The characterization of depth noise, in relation to the particular sensing architecture, is a major issue. This can be addressed using bilateral [49] or nonlocal [18] filters, or in wavelet space [10], using prior knowledge of the spatial noise distribution. Temporal filtering [30] and video-based [8] methods have also been proposed.

The upsampling of low-resolution depth images is another critical issue. One approach is to apply color super-resolution methods on ToF depth images directly [40]. Alternatively, a high-resolution color image can be used as a reference for depth super resolution [1, 48]. The denoising and upsampling problems can also be addressed together [2], and in conjunction with high-resolution monocular [34] or binocular [7] color images.

It is also important to consider the motion artifacts [28] and multipath [13] problems which are characteristic of ToF sensors. The related problem of ToF depth confidence has been addressed using random-forest methods [35]. Other issues with ToF sensors include internal and external calibration [14, 16, 27], as well as range ambiguity [4]. In the case of Kinect, a unified framework of dense depth data extraction and 3D reconstruction has been proposed [33].

Despite the increasing interest in active depth sensors, there are many unresolved issues regarding the data produced by these devices, as outlined above. Furthermore, the lack of any standardized data sets, with ground truth, makes it difficult to make quantitative comparisons between different algorithms.

The Middlebury stereo [38], multiview [41], and Stanford 3D scan [6] data set have been used for the evaluation of depth-image denoising, upsampling, and 3D reconstruction methods. However, these data sets do not provide real depth images taken by either ToF or structured-light depth sensors, and consist of illumination controlled diffuse material objects. While previous depth accuracy enhancement methods demonstrate their experimental results on their own data set, our understanding of the performance and limitations of existing algorithms will remain partial without any quantitative evaluation against a standard data set. This situation hinders the wider adoption and evolution of depth-sensor systems.

In this section, we propose a performance evaluation framework for both ToF and structured-light depth images, based on carefully collected depth maps and their ground truth images. First, we build a standard depth data set; calibrated depth images captured by a ToF depth camera and a structured-light system. Ground truth depth is acquired from a commercial 3D scanner. The data set spans a wide range of objects, organized according to geometric complexity (from smooth to rough), as well as radiometric complexity (diffuse, specular, translucent, and subsurface scattering). We analyze systematic and nonsystematic error sources, including the accuracy and sensitivity with respect to material properties. We also compare the characteristics and performance of the two different types of depth sensors, based on extensive experiments and evaluations. Finally, to justify the usefulness of the data set, we use it to evaluate simple denoising, super resolution, and inpainting algorithms.

1.4.1 Depth Sensors

As described in Sect. 1.2, the ToF depth sensor emits IR waves to target objects, and measures the phase delay of reflected IR waves at each sensor pixel, to calculate the distance traveled. According to the color, reflectivity, and geometric structure of the target object, the reflected IR light shows amplitude and phase variations, causing depth errors. Moreover, the amount of IR is limited by the power consumption of the device, and therefore the reflected IR suffers from low signal-to-noise ratio (SNR). To increase the SNR, ToF sensors bind multiple sensor pixels to calculate a single depth pixel value, which decreases the effective image size. Structured-light depth sensors project an IR pattern onto target objects, which provides a unique illumination code for each surface point observed at by a calibrated IR imaging sensor. Once the correspondence between IR projector and IR sensor is identified by stereo matching methods, the 3D position of each surface point can be calculated by triangulation.

In both sensor types, reflected IR is not a reliable cue for all surface materials. For example, specular materials cause mirror reflection, while translucent materials cause IR refraction. Global illumination also interferes with the IR sensing mechanism, because multiple reflections cannot be handled by either sensor type.

1.4.2 Standard Depth Data Set

A range of commercial ToF depth cameras have been launched in the market, such as PMD, PrimeSense, Fotonic, ZCam, SwissRanger, 3D MLI, and others. Kinect is the first widely successful commercial product to adopt the IR structured-light principle. Among many possibilities, we specifically investigate two depth cameras: a ToF type SR4000 from MESA Imaging [29], and a structured-light type Microsoft Kinect [43]. We select these two cameras to represent each sensor since they are the most popular depth cameras in the research community, accessible in the market and reliable in performance.

Heterogeneous Camera Set

We collect the depth maps of various real objects using the SR4000 and Kinect sensors. To obtain the ground truth depth information, we use a commercial 3D scanning device. As shown in Fig. 1.12, we place the camera set approximately 1.2 m away from the object of interest. The wall behind the object is located about 1.5 m away from the camera set. The specification of each device is as follows.

Mesa SR4000. This is a ToF type depth sensor producing a depth map and amplitude image at the resolution of 176×144 with 16 bit floating-point precision. The amplitude image contains the reflected IR light corresponding to the depth map. In addition to the depth map, it provides $\{x, y, z\}$ coordinates, which correspond to each pixel in the depth map. The operating range of the SR4000 is 0.8–10.0 m, depending on the modulation frequency. The field of view (FOV) of this device is 43×34 degrees.

Kinect. This is a structured IR light type depth sensor, composed of an IR emitter, IR sensor, and color sensor, providing the IR amplitude image, the depth map, and the color image at the resolution of 640×480 (maximum resolution for amplitude and

1.4 Evaluation of Time-of-Flight and Structured-Light Data

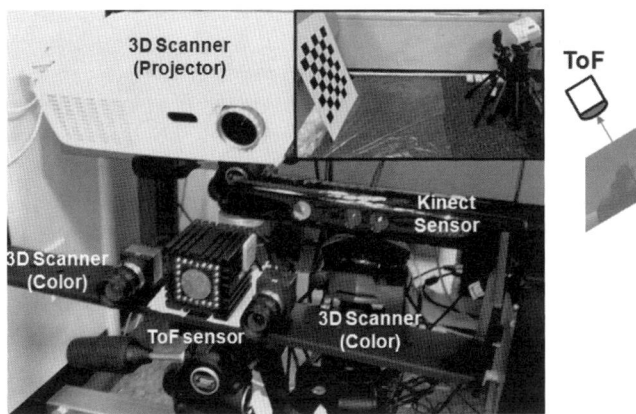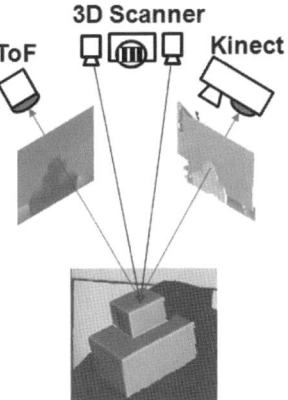

Fig. 1.12 Heterogeneous camera setup for depth sensing

depth image) or 1600×1200 (maximum resolution for RGB image). The operating range is between 0.8 and 3.5 m, the spatial resolution is 3 mm at 2 m distance, and the depth resolution is 10 mm at 2 m distance. The FOV is 57×43 degrees.

FlexScan3D. We use a structured-light 3D scanning system for obtaining ground truth depth. It consists of an LCD projector and two color cameras. The LCD projector illuminates coded pattern at 1024×768 resolution, and each color camera records the illuminated object at 2560×1920 resolution.

Capturing Procedure for Test Images

The important property of the data set is that the measured depth data is aligned with ground truth information, and with that of the other sensor. Each depth sensor has to be fully calibrated internally and externally. We employ a conventional camera calibration method [52] for both depth sensors and the 3D scanner. Intrinsic calibration parameters for the ToF sensors are known. Given the calibration parameters, we can transform ground truth depth maps onto each depth sensor space. Once the system is calibrated, we proceed to capture the objects of interest. For each object, we record depth (**ToFD**) and intensity (**ToFI**) images from the SR4000, plus depth (**SLD**) and color (**SLC**) from the Kinect. Depth captured by the FlexScan3D is used as ground truth (**GTD**), as explained in more detail below (Fig. 1.13).

Data Set

We select objects that show radiometric variations (diffuse, specular, and translucent), as well as geometric variations (smooth or rough). The total 36-item test set is divided

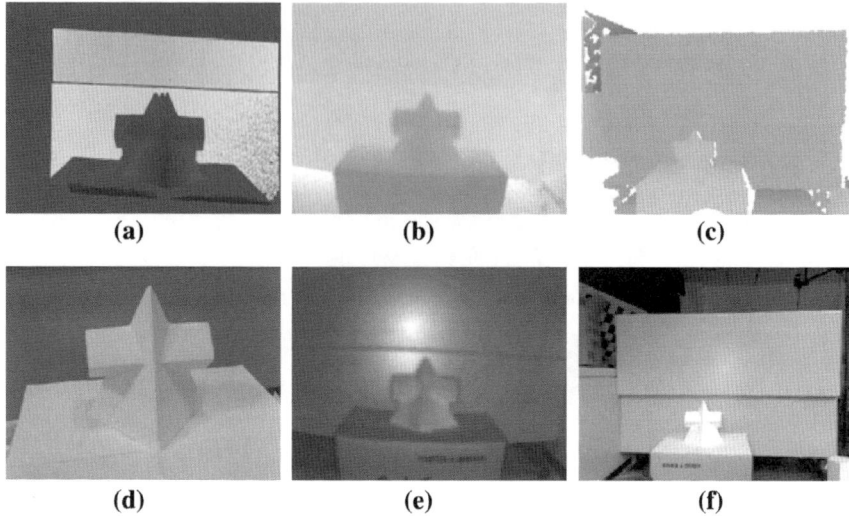

Fig. 1.13 Sample raw image set of depth and ground truth. **a** GTD, **b** ToFD, **c** SLD, **d** Object, **e** ToFI, **f** SLC

into three subcategories: diffuse material objects (class A), specular material objects (class B), and translucent objects with subsurface scattering (class C), as in Fig. 1.15. Each class demonstrates geometric variation from smooth to rough surfaces (a smaller label number means a smoother surface).

From diffuse, through specular to translucent materials, the radiometric representation becomes more complex, requiring a high-dimensional model to predict the appearance. In fact, the radiometric complexity also increases the level of challenges in recovering its depth map. This is because the complex illumination interferes with the sensing mechanism of most depth devices. Hence, we categorize the radiometric complexity by three classes, representing the level of challenges posed by material variation. From smooth to rough surfaces, the geometric complexity is increased, especially due to mesostructure scale variation.

Ground Truth

We use a 3D scanner for ground truth depth acquisition. The principle of this system is similar to [39]; using illumination patterns and solving correspondences and triangulating between matching points to compute the 3D position of each surface point. Simple gray illumination patterns are used, which gives robust performance in practice. However, the patterns cannot be seen clearly enough to provide correspondences for non-Lambertian objects [3]. Recent approaches [15] suggest new high-frequency patterns, and present improvement in recovering depth in the pres-

1.4 Evaluation of Time-of-Flight and Structured-Light Data

Original objects   After matt spray

Fig. 1.14 We apply *white* matt spray on *top* of non-Lambertian objects for ground truth depth acquisition. Original objects after matt spray

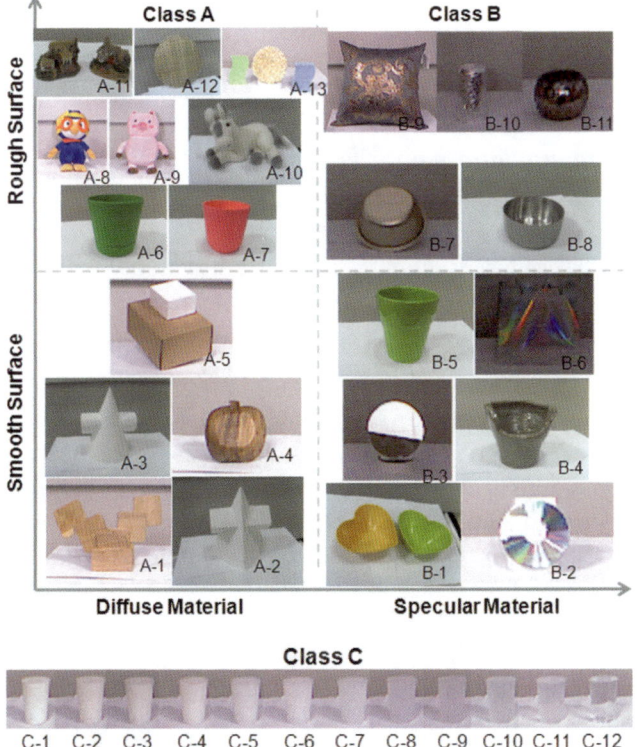

Fig. 1.15 Test images categorized by their radiometric and geometric characteristics: class A diffuse material objects (13 images), class B specular material objects (11 images), and class C translucent objects with subsurface scattering (12 images)

ence of global illumination. Among all surfaces, the performance of structured-light scanning systems is best for Lambertian materials.

The data set includes non-Lambertian materials presenting various illumination effects; specular, translucent, and subsurface scattering. To employ the 3D scanner system for ground truth depth acquisition of the data set, we apply white matt spray on top of each object surface, so that we can give each object a Lambertian surface while we take ground truth depth Fig. 1.14. To make it clear that the spray particles do not change the surface geometry, we have compared the depth maps captured by the 3D scanner before and after the spray on a Lambertian object. We observe that the thickness of spray particles is below the level of the depth-sensing precision, meaning that the spray particles do not affect on the accuracy of the depth map in practice. Using this methodology, we are able to obtain ground truth depth for non-Lambertian objects. To ensure the level of ground truth depth, we capture the depth map of a white board. Then, we apply RANSAC to fit a plane to the depth map and measure the variation of scan data from the plane. We observe that the variation is less than 200 μm, which is negligible compared to depth sensor errors. Finally, we adopt the depth map from the 3D scanner as the ground truth depth, for quantitative evaluation and analysis.

1.4.3 Experiments and Analysis

In this section, we investigate the depth accuracy, the sensitivity to various different materials, and the characteristics of the two types of sensors.

Depth Accuracy and Sensitivity

Given the calibration parameters, we project the ground truth depth map onto each sensor space, in order to achieve viewpoint alignment (Fig. 1.12). Due to the resolution difference, multiple pixels of the ground truth depth fall into each sensor pixel. We perform a bilinear interpolation to find corresponding ground truth depth for each sensor pixel. Due to the difference of field of view and occluded regions, not all sensor pixels get corresponding ground truth depth. We exclude these pixels and occlusion boundaries from the evaluation.

According to previous work [21, 42] and manufacturer reports on the accuracy of depth sensors, the root-mean-square error (RMSE) of depth measurements is approximately 5–20 mm at the distance of 1.5 m. These figures cannot be generalized for all materials, illumination effects, complex geometry, and other factors. The use of more general objects and environmental conditions invariably results in higher RMSE of depth measurement than reported numbers. When we tested with a white wall, which is similar to the calibration object used in previous work [42], we obtain approximately 10.15 mm at the distance of 1.5 m. This is comparable to the previous empirical study and reported numbers.

Because only foreground objects are controlled, the white background is segmented out for the evaluation. The foreground segmentation is straightforward because the background depth is clearly separated from that of foreground. In Figs. 1.16, 1.17 and 1.18, we plot depth errors (RMSE) and show difference maps (8 bit) between the ground truth and depth measurement. In the difference maps, gray indicates zero difference, whereas a darker (or brighter) value indicates that the

1.4 Evaluation of Time-of-Flight and Structured-Light Data

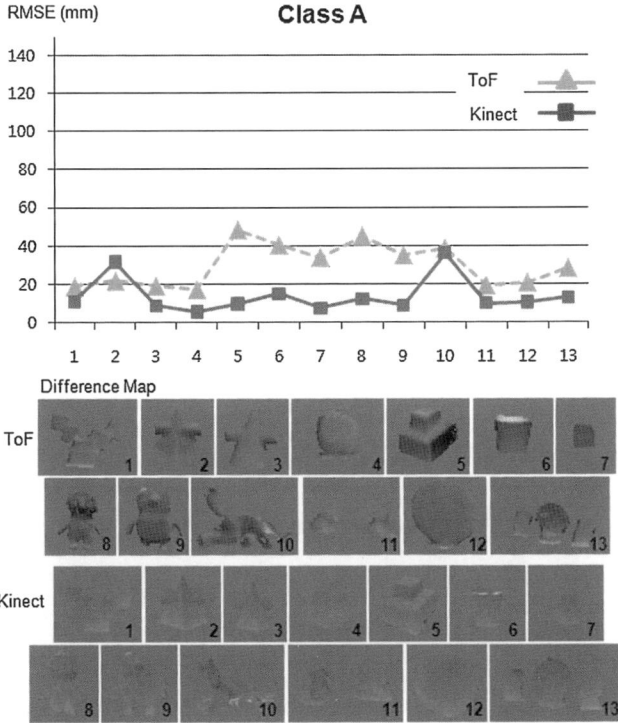

Fig. 1.16 ToF depth accuracy in RMSE (root mean square) for class A. The RMSE values and their corresponding difference maps are illustrated. 128 in difference map represents zero difference while 129 represents the ground truth is 1 mm larger than the measurement. Likewise, 127 indicates that the ground truth is 1 mm smaller than the measurement

ground truth is smaller (or larger) than the estimated depth. The range of difference map, [0, 255], spans [−128 mm, 128 mm] in RMSE.

Several interesting observations can be made from the experiments. First, we observe that the accuracy of depth values varies substantially according to the material property. As shown in Fig. 1.16, the average RMSE of class A is 26.80 mm with 12.81 mm of standard deviation, which is significantly smaller than the overall RMSE. This is expected, because class A has relatively simple properties, which are well approximated by the Lambertian model. From Fig. 1.17 for class B, we are unable to obtain the depth measurements on specular highlights. These highlights either prevent the IR reflection back to the sensor, or cause the reflected IR to saturate the sensor. As a result, the measured depth map shows holes, introducing a large amount of errors. The RMSE for class B is 110.79 mm with 89.07 mm of standard deviation. Class C is the most challenging subset, since it presents the subsurface scattering and translucency. As expected, upon the increase in the level of translucency, the measurement error is dramatically elevated as illustrated in Fig. 1.18.

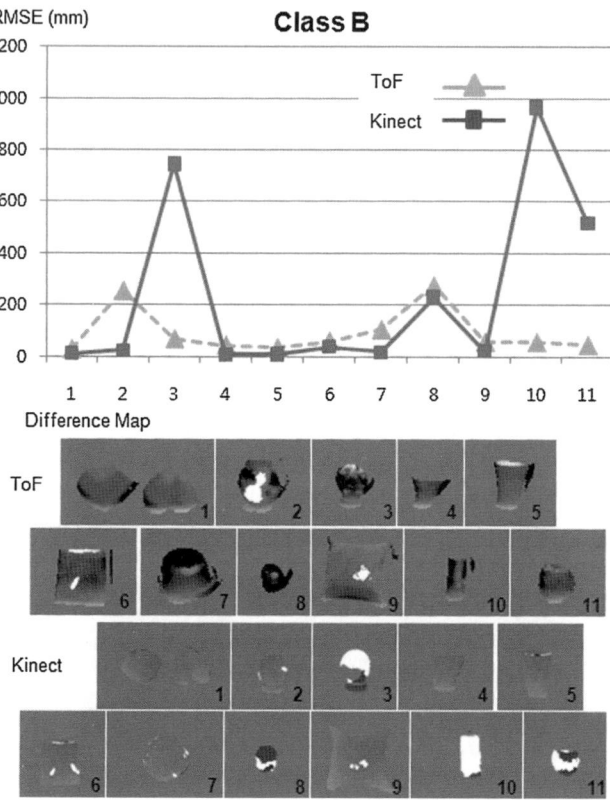

Fig. 1.17 ToF depth accuracy in RMSE (root mean square) for class B. The RMSE values and their corresponding difference maps are illustrated

One thing to note is that the error associated with translucent materials differs from that associated with specular materials. We still observe some depth values for translucent materials, whereas the specular materials show holes in the depth map. The measurement on translucent materials is incorrect, often producing larger depth than the ground truth. Such a drift appears because the depth measurements on translucent materials are the result of both translucent foreground surface and the background behind. As a result, the corresponding measurement points lie somewhere between the foreground and the background surfaces.

Finally, the RMSE for class C is 148.51 mm with 72.19 mm of standard deviation. These experimental results are summarized in Table 1.1. Interestingly, the accuracy is not so much dependent on the geometric complexity of the object. Focusing on class A, although A-11, A-12, and A-13 possess complicated and uneven surface geometry, the actual accuracy is relatively good. Instead, we find that the error increases as the surface normal deviates from the optical axis of the sensor. In fact, a similar problem has been addressed by [22], in that the orientation is the source

1.4 Evaluation of Time-of-Flight and Structured-Light Data

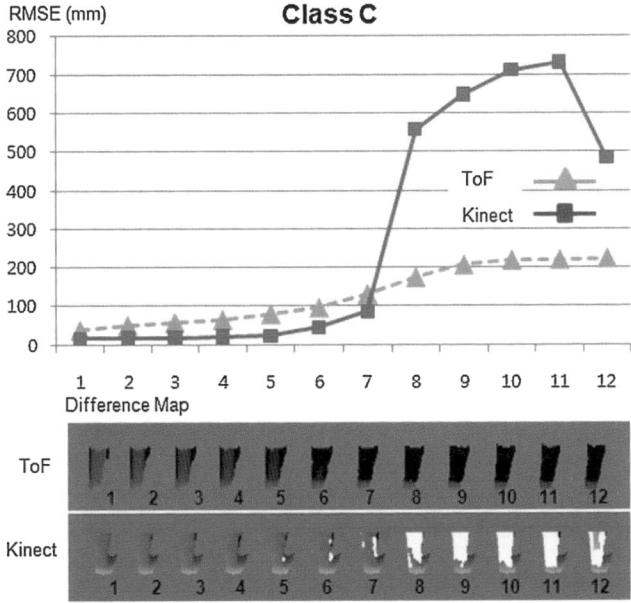

Fig. 1.18 ToF depth accuracy in RMSE (root mean square) for class C. The RMSE values and their corresponding difference maps are illustrated

of systematic error in sensor measurement. In addition, surfaces where the global illumination occurs due to multipath IR transport (such as the concave surfaces on A-5, A-6, A-10 of Class A) exhibit erroneous measurements.

Due to its popular application in games and human computer interaction, many researchers have tested and reported the result of Kinect applications. One of common observation is that the Kinect presents some systematic error with respect to distance. However, there has been no in-depth study on how the Kinect works on various surface materials. We measure the depth accuracy of Kinect using the data set, and illustrate the results in Figs. 1.16, 1.17 and 1.18.

Overall RMSE is 191.69 mm, with 262.19 mm of standard deviation. Although the overall performance is worse than that of ToF sensor, it provides quite accurate results for class A. From the experiments, it is clear that material properties are strongly correlated with depth accuracy. The RMSE for class A is 13.67 mm with 9.25 mm of standard deviation. This is much smaller than the overall RMSE, 212.56 mm. However, the error dramatically increases in class B (303.58 mm with 249.26 mm of deviation). This is because the depth values for specular materials cause holes in the depth map, similar to the behavior of the ToF sensor.

From the experiments on class C, we observe that the depth accuracy drops significantly upon increasing the level of translucency, especially starting at the object C-8. In the graph shown in Fig. 1.18, one can observe that the RMSE is reduced with a completely transparent object (C-12, a pure water). It is because caustic effects

Table 1.1 Depth accuracy upon material properties. Class A: diffuse, Class B: specular, Class C: translucent. See Fig. 1.15 for illustration

	Overall	Class A	Class B	Class C
ToF	83.10	29.68	93.91	131.07
	(76.25)	(10.95)	(87.41)	(73.65)
Kinect	170.153	13.67	235.30	279.96
	(282.25)	(9.25)	(346.44)	(312.97)

Root-mean-square error (standard deviation) in mm

Table 1.2 Depth accuracy before/after bilateral filtering and superresolution for class A. See Fig. 1.19 for illustration

	Original RMSE	Bilateral filtering	Bilinear interpolation
ToF	29.68	27.78	31.93
	(10.95)	(10.37)	(23.34)
Kinect	13.67	13.30	15.02
	(9.25)	(9.05)	(12.61)

Root-mean-square error (standard deviation) in mm

appear along the object, sending back unexpected IR signals to the sensor. Since the sensor receives the reflected IR, RMSE improves in this case. However, this does not always stand for a qualitative improvement. The overall RMSE for class C is 279.96 mm with 312.97 mm of standard deviation. For comparison, see Table 1.1.

ToF Versus Kinect Depth

In previous sections, we have demonstrated the performance of ToF and structured-light sensors. We now characterize the error patterns of each sensor, based on the experimental results. For both sensors, we observe two major errors; data drift and data loss. It is hard to state which kind of error is most serious, but it is clear that both must be addressed. In general, the ToF sensor tends to show data drift, whereas the structured-light sensor suffers from data loss. In particular, the ToF sensor produces a large offset in depth values along boundary pixels and transparent pixels, which correspond to data drift. Under the same conditions, the structured-light sensor tends to produce holes, in which the depth cannot be estimated. For both sensors, specular highlights lead to data loss.

1.4.4 Enhancement

In this section, we apply simple denoising, super resolution and inpainting algorithms on the data set, and report their performance. For denoising and super resolution, we

1.4 Evaluation of Time-of-Flight and Structured-Light Data

Table 1.3 Depth accuracy before/after inpainting for class B. See Fig. 1.20 for illustration

	Original RMSE	Example-based inpainting
ToF	93.91	71.62
	(87.41)	(71.80)
Kinect	235.30	125.73
	(346.44)	(208.66)

Root-mean-square error (standard deviation) in mm

test only on class A, because class B and C often suffer from significant data drift or data loss, which neither denoising nor super resolution alone can address.

By excluding class B and C, it is possible to precisely evaluate the quality gain due to each algorithm. On the other hand, we adopt the image inpainting algorithm on class B, because in this case the typical errors are holes, regardless of sensor type. Although the characteristics of depth images differ from those of color images, we apply color inpainting algorithms on depth images, to compensate for the data loss in class B. We then report the accuracy gain, after filling in the depth holes. Note that the aim of this study is not to claim any state-of-the art technique, but to provide baseline test results on the data set.

We choose a bilateral filter for denoising the depth measurements. The bilateral filter size is set to 3×3 (for ToF, 174×144 resolution) or 10×10 (for Kinect, 640×480 resolution). The standard deviation of the filter is set to 2 in both cases. We compute the RMSE after denoising and obtain 27.78 mm using ToF, and 13.30 mm using Kinect as demonstrated in Tables 1.2 and 1.3. On average, the bilateral filter provides an improvement in depth accuracy; 1.98 mm gain for ToF and 0.37 mm for Kinect. Figure 1.19 shows the noise-removed results, with input depth.

We perform bilinear interpolation for super resolution, increasing the resolution twice per dimension (upsampling by a factor of four). We compute the RMSE before and after the super resolution process from the identical ground truth depth map. The depth accuracy is decreased after super resolution by 2.25 mm (ToF) or 1.35 mm (Kinect). The loss of depth accuracy is expected, because the recovery of surface details from a single low-resolution image is an ill-posed problem. The quantitative evaluation results for denoising and super resolution are summarized in Tables 1.2 and 1.3.

For inpainting, we employ an exemplar-based algorithm [5]. Criminisi et al. designed a fill order to retain the linear structure of scene, and so their method is well suited for depth images. For hole filling, we set the patch size to 3×3 for ToF and to 9×9 for Kinect, in order to account for the difference in resolution. Finally, we compute the RMSE after inpainting, which is 75.71 mm for ToF and 125.73 mm for Kinect. The overall accuracy has been improved by 22.30 mm for ToF and 109.57 mm for Kinect. The improvement for Kinect is more significant than ToF, because the data loss appears more frequently in Kinect than ToF. After the inpainting process, we obtain a reasonable quality improvement for class B.

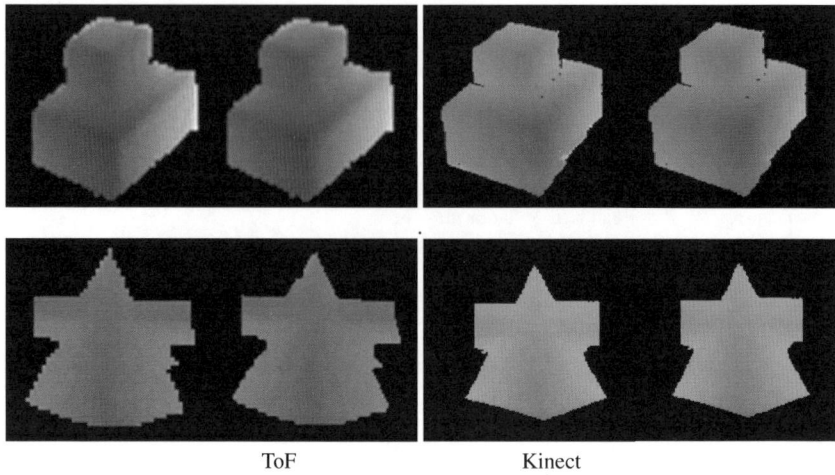

ToF   Kinect

Fig. 1.19 Results before and after bilateral filtering (*top*) and bilinear interpolation (*bottom*)

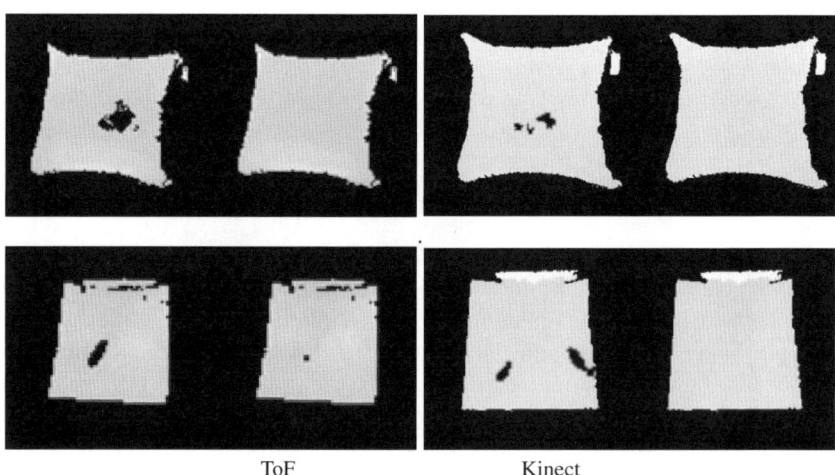

ToF   Kinect

Fig. 1.20 Before and after inpainting

Based on the experimental study, we confirm that both depth sensors provide relatively accurate depth measurements for diffuse materials (class A). For specular materials (class B), both sensors exhibit data loss appearing as holes in the measured depth. Such a data loss causes a large amount of error in the depth images in Fig. 1.21. For translucent materials (class C), the ToF sensor shows nonlinear data drift toward the background. On the other hand, the Kinect sensor shows data loss on translucent materials. Upon the increase of translucency, the performance of both sensors is degraded accordingly.

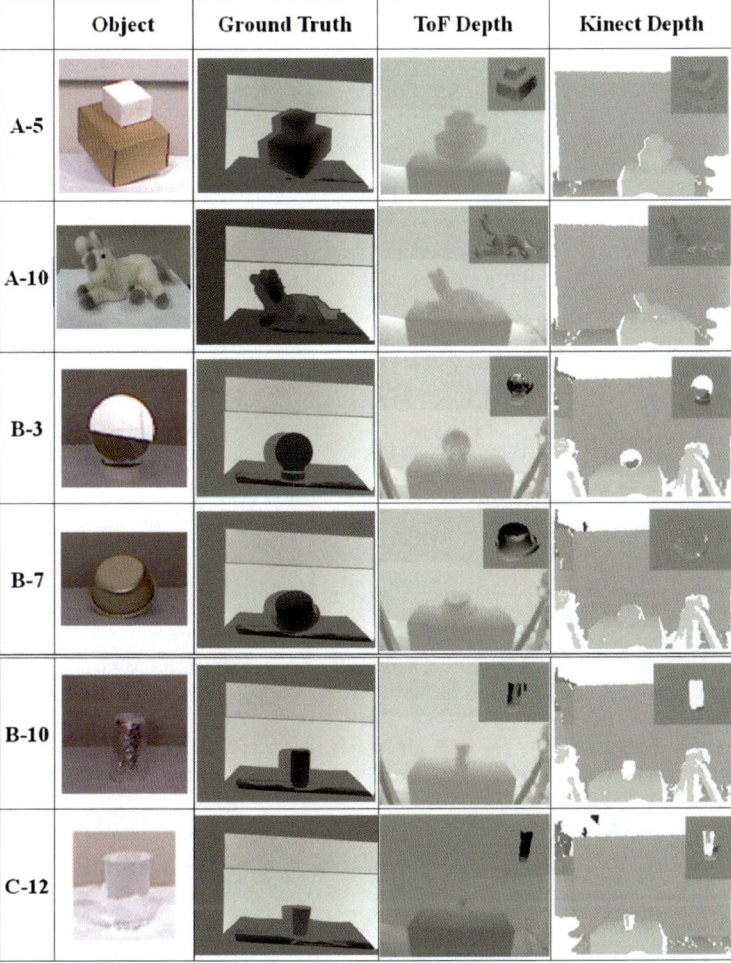

Fig. 1.21 Sample depth images and difference maps from the test image set

1.5 Conclusions

This chapter has reported both quantitative and qualitative experimental results for the evaluation of each sensor type. Moreover, we provide a well-structured standard data set of depth images from real world objects, with accompanying ground truth depth. The data set spans a wide variety of radiometric and geometric complexity, which is well suited to the evaluation of depth processing algorithms. The analysis has revealed important problems in depth acquisition and processing, especially measurement errors due to material properties. The data set will provide a standard framework for the evaluation of other denoising, super resolution, interpolation, and related depth-processing algorithms.

References

1. Bartczak, B., Koch, R.: Dense depth maps from low resolution time-of-flight depth and high resolution color views. In: Proceedings of the International Symposium on Visual Computing (ISVC), pp. 228–239. Las Vegas (2009)
2. Chan, D., Buisman, H., Theobalt, C., Thrun, S.: A noise-aware filter for real-time depth upsampling. In: ECCV Workshop on Multi-camera and Multi-modal Sensor Fusion Algorithms and Applications (2008)
3. Chen, T., Lensch, H.P.A., Fuchs, C., Seidel H.P.: Polarization and phase-shifting for 3D scanning of translucent objects. In: Proceedings of the Computet Vision and, Pattern Recognition, pp. 1–8 (2007)
4. Choi, O., Lim, H., Kang, B., Kim, Y., Lee, K., Kim, J., Kim, C.: Range unfolding for time-of-flight depth cameras. In: Proceedings of the International Conference on Image Processing. pp. 4189–4192 (2010)
5. Criminisi, A., Perez, P., Toyama, K.: Region filling and object removal by exemplar-based image inpainting. IEEE Trans. Image Process. **13**(9), 1200–1212 (2004)
6. Curless, B., Levoy, M.: A volumetric method for building complex models from range images. In: Proceedings of ACM SIGGRAPH '96, pp. 303–312 (1996)
7. Dolson, J., Baek, J., Plagemann, C., Thrun, S.: Fusion of time-of-flight depth and stereo for high accuracy depth maps. In: Proceedings of the Computer Vision and Pattern Pecognition (CVPR), pp. 1–8 (2008)
8. Dolson, J., Baek, J., Plagemann, C., Thrun, S.: Upsampling range data in dynamic environments. In: Proceedings of the Computer Vision and Pattern Recognition (CVPR), pp. 1141–1148 (2010)
9. Du, H., Oggier, T., Lustenberger, F., Charbon, E.: A virtual keyboard based on true-3d optical ranging. In: Proceedings of the British Machine Vision Conference (BMVC'05), pp. 220–229 (2005)
10. Edeler, T., Ohliger, K., Hussmann, S. Mertins, A.: Time-of-flight depth image denoising using prior noise information. In: Proceedings of the IEEE 10th International Conference on Signal Processing (ICSP), pp. 119–122 (2010)
11. Foix, S., Alenya, G., Torras, C.: Lock-in time-of-flight (ToF) cameras: a survey. IEEE Sens. J. **11**(9), 1917–1926 (2011)
12. Freedman, B., Shpunt, A., Machline, M., Arieli, Y.: Depth Mapping Using Projected Patterns. US Patent No. 8150412 (2012)
13. Fuchs, S.: Multipath interference compensation in time-of-flight camera images. In: Proceedings of the 2010 20th International Conference on, Pattern Recognition (ICPR). pp. 3583–3586 (2010)
14. Fuchs, S., Hirzinger, G.: Extrinsic and depth calibration of ToF-cameras. In: Proceedings of the Computer Vision and Pattern Recognition (CVPR), pp. 1–6 (2008)
15. Gupta, M., Agrawal, A., Veeraraghavan, A., Narasimhan, S.G.: Structured light 3D scanning under global illumination. In: Proceedings of the Computer Vision and Pattern Recognition (CVPR) (2011)
16. Hansard, M., Horaud, R., Amat, M., Lee, S.: Projective alignment of range and parallax data. In: Proceedings of the Computer Vision and Pattern Recognition (CVPR), pp. 3089–3096 (2011)
17. Henry, P., Krainin, M., Herbst, E., Ren, X., Fox, D.: RGB-D mapping: using depth cameras for dense 3d modeling of indoor environments. In: RGB-D: Advanced Reasoning with Depth Cameras Workshop in Conjunction with RSS (2010)
18. Huhle, B., Schairer, T., Jenke, P., Strasser. W.: Robust non-local denoising of colored depth data. In: Proceedings of the Computer Vision and Pattern Recognition (CVPR) Workshops, pp. 1–7 (2008)
19. Hussmann, S., Hermanski, A., Edeler, T.: Real-time motion artifact suppression in Tof camera systems. IEEE Trans. Instrum. Meas. **60**(5), 1682–1690 (2011)

20. Kang, B., Kim, S., Lee, S., Lee, K., Kim, J., Kim, C.: Harmonic distortion free distance estimation in Tof camera. In: SPIE Electronic Imaging (2011)
21. Khoshelham, K.: Accuracy analysis of kinect depth data. In: Proceedings of the ISPRS Workshop on Laser Scanning (2011)
22. Kim, Y., Chan, D., Theobalt, C., Thrun, S.: Design and calibration of a multi-view TOF sensor fusion system. In: Proceedings of the IEEE CVPR Workshop on Time-of-Flight Camera Based Computer Vision (2008)
23. Kolb, A., Barth, E., Koch, R., Larsen, R.: Time-of-flight cameras in computer graphics. Comput. Graph Forum **29**(1), 141–159 (2010)
24. Lee, S., Kang, B., Kim, J.D.K., Kim, C.-Y.: Motion blur-free time-of-flight range sensor. In: Proceedings of the SPIE Electronic Imaging (2012)
25. Lee, S., Shim, H., Kim, J.D.K., Kim, C.-Y.: Tof depth image motion blur detection using 3D blur shape models. In: Proceedings of the SPIE Electronic Imaging (2012)
26. Lindner, M., Kolb, A.: Compensation of motion artifacts for time-of-flight cameras. In: Kolb, A., Koch, R. (eds.) Dynamic 3D Imaging, Lecture Notes in Computer Science, vol. 5742, pp. 16–27. Springer, Berlin (2009)
27. Lindner, M., Kolb, A., Ringbeck, T.: New insights into the calibration of tof-sensors. In: Proceedings on Computer Vision and Pattern Recognition Workshops, pp. 1–5 (2008)
28. Lottner, O., Sluiter, A., Hartmann, K., Weihs, W.: Movement artefacts in range images of time-of-flight cameras. In: International Symposium on Signals, Circuits and Systems (ISSCS), vol. 1, pp. 1–4 (2007)
29. Mesa Imaging AG. http://www.mesa-imaging.ch
30. Matyunin, S., Vatolin, D., Berdnikov, Y., Smirnov, M.: Temporal filtering for depth maps generated by kinect depth camera. In: Proceedings of the 3DTV, pp. 1–4 (2011)
31. May, S., Werner, B., Surmann, H., Pervolz, K.: 3D Time-of-flight cameras for mobile robotics. In: Proceedings of IEEE/RSJ International Conference on Intelligent Robots and Systems, pp. 790–795 (2006)
32. Mure-Dubois, J., Hugli, H.: Real-time scattering compensation for time-of-flight camera. In: Proceedings of Workshop on Camera Calibration Methods for Computer Vision Systems (CCMVS2007) (2007)
33. Newcombe, R.A., Izadi, S., Hilliges, O., Molyneaux, D., Kim, D., Davison, A.J., Kohli, P., Shotton, J., Hodges, S., Fitzgibbon, A.: Kinectfusion: real-time dense surface mapping and tracking. IEEE International Symposium on Mixed and Augmented Reality (ISMAR), pp. 1–8 (2011)
34. Park, J., Kim, H., Tai, Y.-W., Brown, M.-S., Kweon, I.S.: High quality depth map upsampling for 3D-TOF cameras. In: Proceedings of IEEE International Conference on Computer Vision (ICCV) (2011)
35. Reynolds, M., Dobos, J., Peel, L., Weyrich, T., Brostow, G.: Capturing time-of-flight data with confidence. In: Proceedings of the Computer Vision and Pattern Recognition (CVPR), pp. 945–952 (2011)
36. Ryden, F., Chizeck, H., Kosari, S.N., King, H., Hannaford, B.: Using kinect and a haptic interface for implementation of real-time virtual fixtures. In: RGB-D: Advanced Reasoning with Depth Cameras Workshop in Conjunction with RSS (2010)
37. Schamm, T., Strand, M., Gumpp, T., Kohlhaas, R., Zollner, J., Dillmann, R.: Vision and Tof-based driving assistance for a personal transporter. In: Proceedings of the International Conference on Advanced Robotics (ICAR), pp. 1–6 (2009)
38. Scharstein, D., Szeliski, R.: A taxonomy and evaluation of dense two-frame stereo correspondence algorithms. Int. J. Comput. Vision **47**, 7–42 (2002)
39. Scharstein, D., Szeliski, R.: High-accuracy stereo depth maps using structured light. In: Proceedings of the Computer Vision and Pattern Recognition (CVPR) (2003)
40. Schuon, S., Theobalt, C., Davis, J., Thrun, S.: High-quality scanning using time-of-flight depth superresolution. In: Proceedings of the Computer Vision and Pattern Recognition Workshops, pp. 1–7 (2008)

41. Seitz, S., Curless, B., Diebel, J., Scharstein, D., Szeliski, R.: A comparison and evaluation of multi-view stereo reconstruction algorithms. In: Proceedings of the Computer Vision and Pattern Recognition (CVPR), pp. 519–528 (2006)
42. Shim, H., Adels, R., Kim, J., Rhee, S., Rhee, T., Kim, C., Sim, J., Gross, M.: Time-of-flight sensor and color camera calibration for multi-view acquisition. In: The Visual Computer (2011)
43. Shotton, J., Fitzgibbon, A., Cook, M., Blake, A.: Real-time human pose recognition in parts from single depth images. In: Proceedings of the Computer Vision and Pattern Recognition (CVPR) (2011)
44. Smisek, J., Jancosek, M., Pajdla, T.: 3D with kinect. In: Proceedings of International Conference on Computer Vision Workshops, pp. 1154–1160 (2011)
45. Soutschek, S., Penne, J., Hornegger, J., Kornhuber, J.: 3-D Gesture-based scene navigation in medical imaging applications using time-of-flight cameras. In: Proceedings of the Computer Vision and Pattern Recognition Workshops, pp. 1–6 (2008)
46. Tai, Y.-W., Kong, N., Lin, S., Shin, S.Y.: Coded exposure imaging for projective motion deblurring. In: Proceedings of the Computer Vision and Pattern Recognition (CVPR), pp. 2408–2415 (2010)
47. Whyte, O., Sivic, J., Zisserman, A., Ponce, J.: Non-uniform deblurring for shaken images. In: Proceedings of the Computer Vision and Pattern Recognition (CVPR), pp. 491–498 (2010)
48. Yang, Q., Yang, R., Davis, J., Nister, D.: Spatial-depth super resolution for range images. In: Proceedings of the Computer Vision and Pattern Recognition (CVPR), pp. 1–8 (2007)
49. Yeo, D., ul Haq, E., Kim, J., Baig, M., Shin, H.: Adaptive bilateral filtering for noise removal in depth upsampling. In: International SoC Design Conference (ISOCC), pp. 36–39 (2011)
50. Yuan, F., Swadzba, A., Philippsen, R., Engin, O., Hanheide, M., Wachsmuth, S.: Laser-based navigation enhanced with 3D time-of-flight data. In: Proceedings of the International Conference on Robotics and Automation (ICRA'09), pp. 2844–2850 (2009)
51. Zhang, L., Deshpande, A., Chen, X.: Denoising vs. deblurring: hdr imaging techniques using moving cameras. In: Proceedings of the Computer Vision and Pattern Recognition (CVPR), pp. 522–529 (2010)
52. Zhang, Z.: Flexible camera calibration by viewing a plane from unknown orientations. In: Proceedings of the International Conference on Computer Vision (ICCV) (1999)

Chapter 2
Disambiguation of Time-of-Flight Data

Abstract The maximum range of a time-of-flight camera is limited by the periodicity of the measured signal. Beyond a certain range, which is determined by the signal frequency, the measurements are confounded by phase wrapping. This effect is demonstrated in real examples. Several phase-unwrapping methods, which can be used to extend the range of time-of-flight cameras, are discussed. Simple methods can be based on the measured amplitude of the reflected signal, which is itself related to the depth of objects in the scene. More sophisticated unwrapping methods are based on zero-curl constraints, which enforce spatial consistency on the phase measurements. Alternatively, if more than one depth camera is used, then the data can be unwrapped by enforcing consistency among different views of the same scene point. The relative merits and shortcomings of these methods are evaluated, and the prospects for hardware-based approaches, involving frequency modulation are discussed.

Keywords Time-of-Flight principle · Depth ambiguity · Phase unwrapping · Multiple depth cameras

2.1 Introduction

Time-of-Flight cameras emit modulated infrared light and detect its reflection from the illuminated scene points. According to the TOF principle described in Chap. 1, the detected signal is gated and integrated using internal reference signals, to form the tangent of the phase ϕ of the detected signal. Since the tangent of ϕ is a periodic function with a period of 2π, the value $\phi + 2n\pi$ gives exactly the same tangent value for any nonnegative integer n.

Commercially available TOF cameras compute ϕ on the assumption that ϕ is within the range of $[0, 2\pi)$. For this reason, each modulation frequency f has its maximum range $d_{\max}$ corresponding to 2π, encoded without ambiguity:

$$d_{\max} = \frac{c}{2f}, \qquad (2.1)$$

where c is the speed of light. For any scene points farther than $d_{\max}$, the measured distance d is much shorter than its actual distance $d + nd_{\max}$. This phenomenon is called *phase wrapping*, and estimating the unknown number of wrappings n is called *phase unwrapping*.

For example, the Mesa SR4000 [16] camera records a 3D point $\mathbf{X}_p$ at each pixel p, where the measured distance d_p equals $\|\mathbf{X}_p\|$. In this case, the unwrapped 3D point $\mathbf{X}_p(n_p)$ with number of wrappings n_p can be written as

$$\mathbf{X}_p(n_p) = \frac{d_p + n_p d_{\max}}{d_p} \mathbf{X}_p. \qquad (2.2)$$

Figure 2.1a shows a typical depth map acquired by the SR4000 [16], and Fig. 2.1b shows its unwrapped depth map. As shown in Fig. 2.1e, phase unwrapping is crucial for recovering large-scale scene structure.

To increase the usable range of ToF cameras, it is also possible to extend the maximum range $d_{\max}$ by decreasing the modulation frequency f. In this case, the integration time should also be extended, to acquire a high quality depth map, since the depth noise is inversely proportional to f. With extended integration time, moving objects are more likely to result in motion artifacts. In addition, we do not know at which modulation frequency phase wrapping does not occur, without exact knowledge regarding the scale of the scene.

If we can accurately unwrap a depth map acquired at a high modulation frequency, then the unwrapped depth map will suffer less from noise than a depth map acquired at a lower modulation frequency, integrated for the same time. Also, if a phase-unwrapping method does not require exact knowledge on the scale of the scene, then the method will be applicable in more large-scale environments.

There exist a number of phase-unwrapping methods [4–8, 14, 17, 21] that have been developed for ToF cameras. According to the number of input depth maps, the methods are categorized into two groups: those using a single depth map [5, 7, 14, 17, 21] and those using multiple depth maps [4, 6, 8, 20]. The following subsections introduce their principles, advantages and limitations.

2.2 Phase Unwrapping from a Single Depth Map

ToF cameras such as the SR4000 [16] provide an amplitude image along with its corresponding depth map. The amplitude image is encoded with the strength of the detected signal, which is inversely proportional to the squared distance. To obtain *corrected amplitude A'* [19], which is proportional to the reflectivity of a scene surface with respect to the infrared light, we can multiply amplitude A and its corresponding squared distance d^2:

2.2 Phase Unwrapping from a Single Depth Map

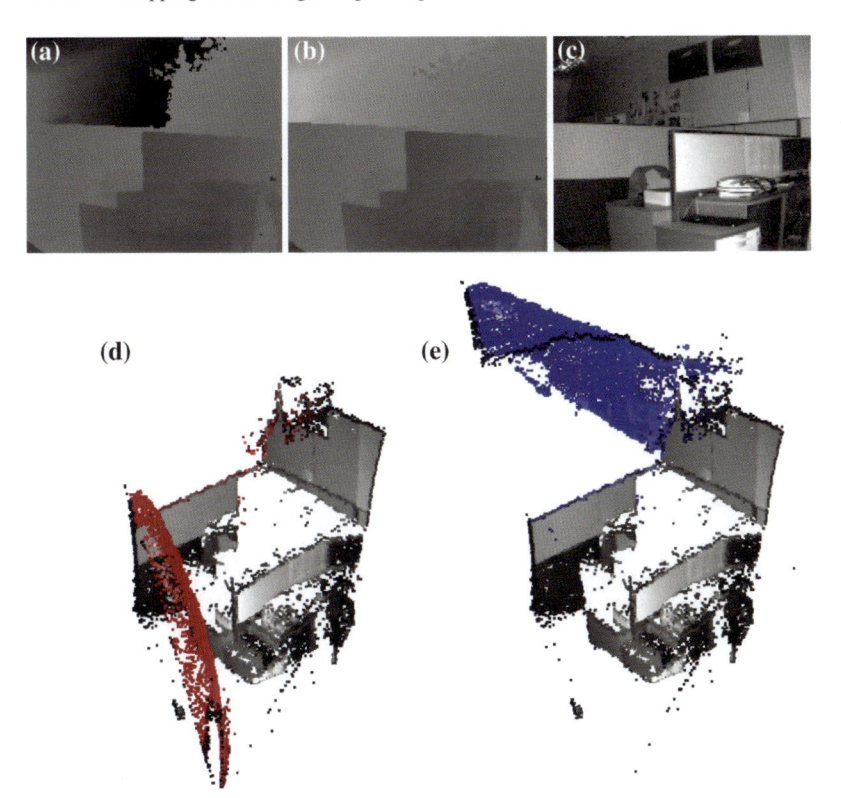

Fig. 2.1 Structure recovery through phase unwrapping. **a** Wrapped ToF depth map. **b** Unwrapped depth map corresponding to (a). Only the distance values are displayed in (a) and (b), to aid visibility. The intensity is proportional to the distance. **c** Amplitude image associated with (a). **d** and **e** display the 3D points corresponding to (a) and (b), respectively. **d** The wrapped points are displayed in *red*. **e** Their unwrapped points are displayed in *blue*. The remaining points are textured using the original amplitude image (c)

$$A' = Ad^2. \tag{2.3}$$

Figure 2.2 shows an example of amplitude correction. It can be observed from Fig. 2.2c that the corrected amplitude is low in the wrapped region. Based on the assumption that the reflectivity is constant over the scene, the corrected amplitude values can play an important role in detecting wrapped regions [5, 17, 21].

Poppinga and Birk [21] use the following inequality for testing if the depth of pixel p has been wrapped:

$$A'_p \leq A^{\text{ref}}_p T, \tag{2.4}$$

where T is a manually chosen threshold, and A^{ref}_p is the reference amplitude of pixel p when viewing a white wall at 1 m, approximated by

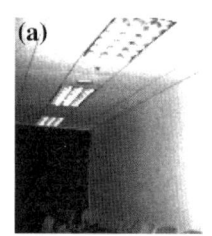

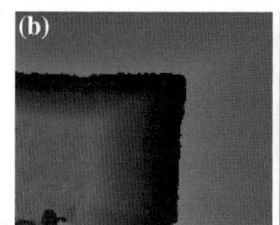

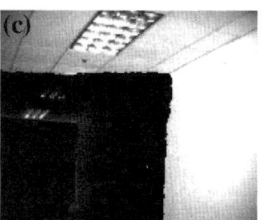

Fig. 2.2 Amplitude correction example. **a** Amplitude image. **b** ToF depth map. **c** Corrected amplitude image. The intensity in (b) is proportional to the distance. The lower left part of (b) has been wrapped. Images courtesy of Choi et al. [5]

$$A_p^{\text{ref}} = B - \left((x_p - c_x)^2 + (y_p - c_y)^2\right), \tag{2.5}$$

where B is a constant. The image coordinates of p are (x_p, y_p), and (c_x, c_y) is approximately the image center, which is usually better illuminated than the periphery. A_p^{ref} compensates this effect by decreasing $A_p^{\text{ref}} T$ if pixel p is in the periphery.

After the detection of wrapped pixels, it is possible to directly obtain an unwrapped depth map by setting the number of wrappings of the wrapped pixels to one on the assumption that the maximum number of wrappings is 1.

The assumption on the constant reflectivity tends to be broken when the scene is composed of different objects with varying reflectivity. This assumption cannot be fully relaxed without detailed knowledge of scene reflectivity, which is hard to obtain in practice. To robustly handle varying reflectivity, it is possible to adaptively set the threshold for each image and to enforce spatial smoothness on the detection results.

Choi et al. [5] model the distribution of corrected amplitude values in an image using a mixture of Gaussians with two components, and apply expectation maximization [1] to learn the model:

$$p(A'_p) = \alpha_H p(A'_p | \mu_H, \sigma_H^2) + \alpha_L p(A'_p | \mu_L, \sigma_L^2), \tag{2.6}$$

where $p(A'_p | \mu, \sigma^2)$ denotes a Gaussian distribution with mean μ and variance σ^2, and α is the coefficient for each distribution. The components $p(A'_p | \mu_H, \sigma_H^2)$ and $p(A'_p | \mu_L, \sigma_L^2)$ describe the distributions of high and low corrected amplitude values, respectively. Similarly, the subscripts H and L denote labels *high* and *low*, respectively. Using the learned distribution, it is possible to write a probabilistic version of Eq. (2.4) as

$$P(H|A'_p) < 0.5, \tag{2.7}$$

where $P(H|A'_p) = \alpha_H p(A'_p | \mu_H, \sigma_H^2) / p(A'_p)$.

To enforce spatial smoothness on the detection results, Choi et al. [5] use a segmentation method [22] based on Markov random fields (MRFs). The method finds the binary labels $n \in \{H, L\}$ or $\{0, 1\}$ that minimize the following energy:

2.2 Phase Unwrapping from a Single Depth Map

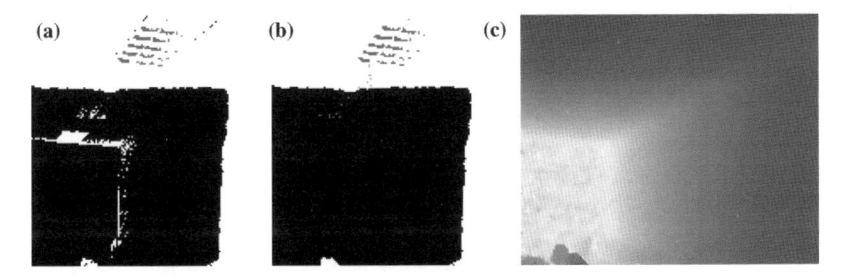

Fig. 2.3 Detection of wrapped regions. **a** Result obtained by expectation maximization. **b** Result obtained by MRF optimization. The pixels with labels L and H are colored in *black* and *white*, respectively. The *red* pixels are those with extremely high or low amplitude values, which are not processed during the classification. **c** Unwrapped depth map corresponding to Fig. 2.2(b). The intensity is proportional to the distance. Images courtesy of Choi et al. [5]

$$E = \sum_p D_p(n_p) + \sum_{(p,q)} V(n_p, n_q), \tag{2.8}$$

where $D_p(n_p)$ is a data cost that is defined as $1 - P(n_p|A'_p)$, and $V(n_p, n_q)$ is a discontinuity cost that penalizes a pair of adjacent pixels p and q if their labels n_p and n_q are different. $V(n_p, n_q)$ is defined in a manner of increasing the penalty if a pair of adjacent pixels have similar corrected amplitude values:

$$V(n_p, n_q) = \lambda \exp\left(-\beta(A'_p - A'_q)^2\right) \delta(n_p \neq n_q), \tag{2.9}$$

where λ and β are constants, which are either manually chosen or adaptively determined. $\delta(x)$ is a function that evaluates to 1 if its argument is true and evaluates to zero otherwise.

Figure 2.3 shows the classification results obtained by Choi et al. [5] Because of varying reflectivity of the scene, the result in Fig. 2.3a exhibits misclassified pixels in the lower left part. The misclassification is reduced by applying the MRF optimization as shown in Fig. 2.3b. Figure 2.3c shows the unwrapped depth map obtained by Choi et al. [5], corresponding to Fig. 2.2b.

McClure et al. [17] also use a segmentation-based approach, in which the depth map is segmented into regions by applying the watershed transform [18]. In their method, wrapped regions are detected by checking the average corrected amplitude of each region.

On the other hand, depth values tend to be highly discontinuous across the wrapping boundaries, where there are transitions in the number of wrappings. For example, the depth maps in Figs. 2.1a, 2.2b shows such discontinuities. On the assumption that the illuminated surface is smooth, the depth difference between adjacent pixels should be small. If the difference between measured distances is greater than $0.5d_{\max}$ for any adjacent pixels, say $d_p - d_q > 0.5d_{\max}$, we can set the number of relative wrappings, or, briefly, the shift $n_q - n_p$ to 1 so that the unwrapped difference will satisfy $-0.5d_{\max} \leq d_p - d_q - (n_q - n_p)d_{\max} < 0$, minimizing the discontinuity.

0.6	0.8	0.9	0.1	0.2	0.3	0.6	0.8	0.9	1.1	1.2	1.3
	p			q			p			q	

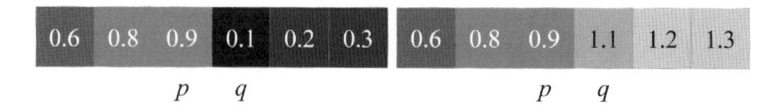

Fig. 2.4 One-dimensional phase-unwrapping example. **a** Measured phase image. **b** Unwrapped phase image where the phase difference between p and q is now *less* than 0.5. In (a) and (b), all the phase values have been divided by 2π. For example, the displayed value 0.1 corresponds to 0.2π

Fig. 2.5 Two-dimensional phase-unwrapping example. **a** Measured phase image. (**b**–**d**) Sequentially unwrapped phase images where the phase difference across the *red dotted line* has been minimized. From **a** to **d**, all the phase values have been divided by 2π. For example, the displayed value 0.1 corresponds to 0.2π

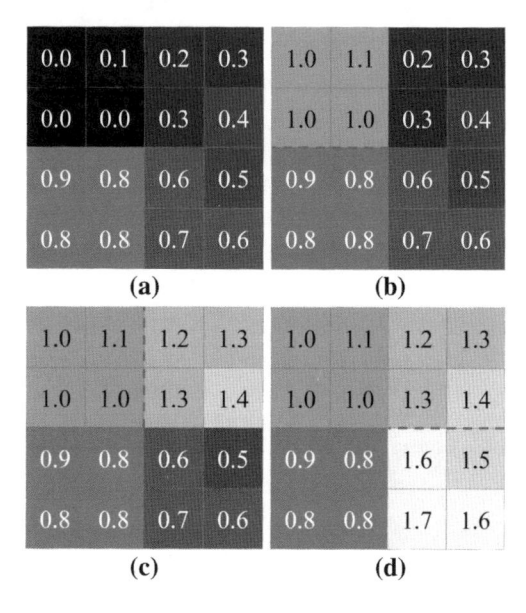

(a)
0.0	0.1	0.2	0.3
0.0	0.0	0.3	0.4
0.9	0.8	0.6	0.5
0.8	0.8	0.7	0.6

(b)
1.0	1.1	0.2	0.3
1.0	1.0	0.3	0.4
0.9	0.8	0.6	0.5
0.8	0.8	0.7	0.6

(c)
1.0	1.1	1.2	1.3
1.0	1.0	1.3	1.4
0.9	0.8	0.6	0.5
0.8	0.8	0.7	0.6

(d)
1.0	1.1	1.2	1.3
1.0	1.0	1.3	1.4
0.9	0.8	1.6	1.5
0.8	0.8	1.7	1.6

Figure 2.4 shows a one-dimensional phase-unwrapping example. In Fig. 2.4a, the phase difference between pixels p and q is greater than 0.5 (or π). The shifts that minimize the difference between adjacent pixels are 1 (or, $n_q - n_p = 1$) for p and q, and 0 for the other pairs of adjacent pixels. On the assumption that n_p equals 0, we can integrate the shifts from left to right to obtain the unwrapped phase image in Fig. 2.4b.

Figure 2.5 shows a two-dimensional phase-unwrapping example. From Fig. 2.5a to d, the phase values are unwrapped in a manner of minimizing the phase difference across the red dotted line. In this two-dimensional case, the phase differences greater than 0.5 never vanish, and the red dotted line cycles around the image center infinitely. This is because of the local phase error that causes the violation of the *zero-curl* constraint [9, 12].

Figure 2.6 illustrates the zero-curl constraint. Given four neighboring pixel locations (x, y), $(x + 1, y)$, $(x, y + 1)$, and $(x + 1, y + 1)$, let $a(x, y)$ and $b(x, y)$ denote the shifts $n(x + 1, y) - n(x, y)$ and $n(x, y + 1) - n(x, y)$, respectively, where $n(x, y)$ denotes the number of wrappings at (x, y). Then, the shift $n(x+1, y+1) - n(x, y)$ can be calculated in two different ways: either $a(x, y) + b(x+1, y)$ or $b(x, y) + a(x, y+1)$

2.2 Phase Unwrapping from a Single Depth Map

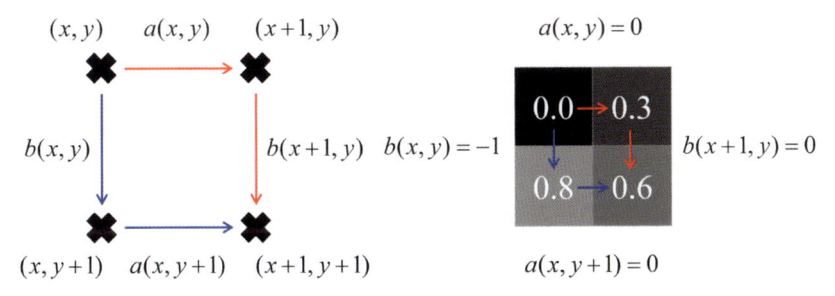

Fig. 2.6 Zero-curl constraint: $a(x, y) + b(x + 1, y) = b(x, y) + a(x, y + 1)$. **a** The number of relative wrappings between $(x+1, y+1)$ and (x, y) should be consistent regardless of its integrating paths. For example, two different paths (*red* and *blue*) are shown. **b** shows an example in which the constraint is not satisfied. The *four* pixels correspond to the *four* pixels in the middle of Fig. 2.5a

following one of the two different paths shown in Fig. 2.6a. For any phase-unwrapping results to be consistent, the two values should be the same, satisfying the following equality:

$$a(x, y) + b(x + 1, y) = b(x, y) + a(x, y + 1). \tag{2.10}$$

Because of noise or discontinuities in the scene, the zero-curl constraint may not be satisfied locally, and the local error is propagated to the entire image during the integration. There exist classical phase-unwrapping methods [9, 12] applied in magnetic resonance imaging [15] and interferometric synthetic aperture radar (SAR) [13], which rely on detecting [12] or fixing [9] broken zero-curl constraints. Indeed, these classical methods [9, 12] have been applied to phase unwrapping for ToF cameras [7, 14].

2.2.1 Deterministic Methods

Goldstein et al. [12] assume that the shift is either 1 or -1 between adjacent pixels if their phase difference is greater than π, and assume that it is 0 otherwise. They detect cycles of four neighboring pixels, referred to as *plus and minus residues*, which do not satisfy the zero-curl constraint.

If any integration path encloses an unequal number of plus and minus residue, the integrated phase values on the path suffer from global errors. In contrast, if any integration path encloses an equal number of plus and minus residues, the global error is balanced out. To prevent global errors from being generated, Goldstein et al. [12] connect nearby plus and minus residues with *cuts*, which interdict the integration paths, such that no net residues can be encircled.

After constructing the cuts, the integration starts from a pixel p, and each neighboring pixel q is unwrapped relatively to p in a greedy and sequential manner if q has not been unwrapped and if p and q are on the same side of the cuts.

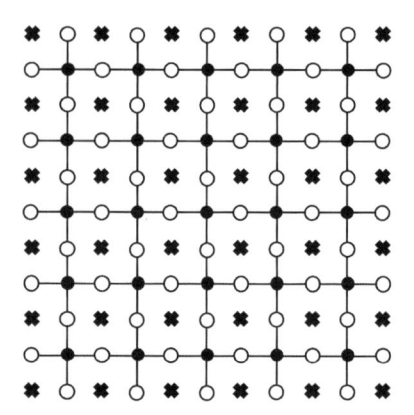

Fig. 2.7 Graphical model that describes the zero-curl constraints (*black discs*) between neighboring shift variables (*white discs*). 3-element probability vectors (μ's) on the shifts between adjacent nodes (-1, 0, or 1) are propagated across the network. The x marks denote pixels [9]

2.2.2 Probabilistic Methods

Frey et al. [9] propose a very loopy belief propagation method for estimating the shift that satisfies the zero-curl constraints. Let the set of shifts, and a measured phase image, be denoted by

$$S = \left\{ a(x, y), \, b(x, y) \, : \, x = 1, \ldots, N-1; \, y = 1, \ldots, M-1 \right\}$$

and

$$\Phi = \left\{ \phi(x, y) \, : \, 0 \leq \phi(x, y) < 1, \, x = 1, \ldots, N; \, y = 1, \ldots, M \right\},$$

respectively, where the phase values have been divided by 2π. The estimation is then recast as finding the solution that maximizes the following joint distribution:

$$\begin{aligned}
p(S, \Phi) \propto & \prod_{x=1}^{N-1} \prod_{y=1}^{M-1} \delta(a(x, y) + b(x+1, y) - a(x, y+1) - b(x, y)) \\
& \times \prod_{x=1}^{N-1} \prod_{y=1}^{M} e^{-(\phi(x+1,y) - \phi(x,y) + a(x,y))^2 / 2\sigma^2} \\
& \times \prod_{x=1}^{N} \prod_{y=1}^{M-1} e^{-(\phi(x,y+1) - \phi(x,y) + b(x,y))^2 / 2\sigma^2}
\end{aligned}$$

where $\delta(x)$ evaluates to 1 if $x = 0$ and to 0 otherwise. The variance σ^2 is estimated directly from the wrapped phase image [9].

2.2 Phase Unwrapping from a Single Depth Map

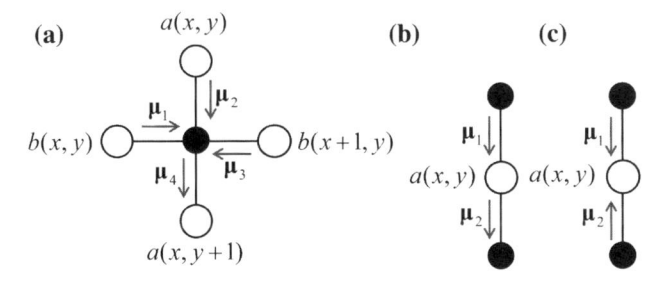

Fig. 2.8 **a** Constraint-to-shift vectors are computed from incoming shift-to-constraint vectors. **b** Shift-to-constraint vectors are computed from incoming constraint-to-shift vectors. **c** Estimates of the marginal probabilities of the shifts given the data are computed by combining incoming constraint-to-shift vectors [9]

Frey et al. [9] construct a graphical model describing the factorization of $p(S, \Phi)$, as shown in Fig. 2.7. In the graph, each shift node (white disc) is located between two pixels, and corresponds to either an x-directional shift (a's) or a y-directional shift (b's). Each constraint node (black disc) corresponds to a zero-curl constraint, and is connected to its four neighboring shift nodes. Every node passes a message to its neighboring node, and each message is a 3-vector denoted by μ, whose elements correspond to the allowed values of shifts, -1, 0, and 1. Each element of μ can be considered as a probability distribution over the three possible values [9].

Figure 2.8a illustrates the computation of a message μ_4 from a constraint node to one of its neighboring shift nodes. The constraint node receives messages μ_1, μ_2, and μ_3 from the rest of its neighboring shift nodes, and filters out the joint message elements that do not satisfy the zero-curl constraint:

$$\mu_{4i} = \sum_{j=-1}^{1} \sum_{k=-1}^{1} \sum_{l=-1}^{1} \delta(k+l-i-j) \mu_{1j} \mu_{2k} \mu_{3l}, \qquad (2.11)$$

where μ_{4i} denotes the element of μ_4, corresponding to shift value $i \in \{-1, 0, 1\}$.

Figure 2.8b illustrates the computation of a message μ_2 from a shift node to one of its neighboring constraint node. Among the elements of the message μ_1 from the other neighboring constraint node, the element, which is consistent with the measured shift $\phi(x, y) - \phi(x+1, y)$, is amplified:

$$\mu_{2i} = \mu_{1i} \exp\left(-\left(\phi(x+1, y) - \phi(x, y) + i\right)^2 / 2\sigma^2\right). \qquad (2.12)$$

After the messages converge (or, after a fixed number of iterations), an estimate of the marginal probability of a shift is computed by using the messages passed into its corresponding shift node, as illustrated in Fig. 2.8c:

$$\hat{P}\big(a(x,y) = i|\Phi\big) = \frac{\mu_{1i}\mu_{2i}}{\sum_j \mu_{1j}\mu_{2j}}. \qquad (2.13)$$

Given the estimates of the marginal probabilities, the most probable value of each shift node is selected. If some zero-curl constraints remain violated, a robust integration technique, such as least-squares integration [10] should be used [9].

2.2.3 Discussion

The aforementioned phase-unwrapping methods using a single depth map [5, 7, 14, 17, 21] have an advantage that the acquisition time is not extended, keeping the motion artifacts at a minimum. The methods, however, rely on strong assumptions that are fragile in real world situations. For example, the reflectivity of the scene surface may vary in a wide range. In this case, it is hard to detect wrapped regions based on the corrected amplitude values. In addition, the scene may be discontinuous if it contains multiple objects that occlude one another. In this case, the wrapping boundaries tend to coincide with object boundaries, and it is often hard to observe large depth discontinuities across the boundaries, which play an important role in determining the number of relative wrappings.

The assumptions can be relaxed by using multiple depth maps at a possible extension of acquisition time. The next subsection introduces phase-unwrapping methods using multiple depth maps.

2.3 Phase Unwrapping from Multiple Depth Maps

Suppose that a pair of depth maps M_1 and M_2 of a static scene are given, which have been taken at different modulation frequencies f_1 and f_2 from the same viewpoint. In this case, pixel p in M_1 corresponds to pixel p in M_2, since the corresponding region of the scene is projected onto the same location of M_1 and M_2. Thus, the unwrapped distances at those corresponding pixels should be consistent within the noise level.

Without prior knowledge, the noise in the unwrapped distance can be assumed to follow a zero-mean distribution. Under this assumption, the maximum likelihood estimates of the numbers of wrappings at the corresponding pixels should minimize the difference between their unwrapped distances. Let m_p and n_p be the numbers of wrappings at pixel p in M_1 and M_2, respectively. Then, we can choose m_p and n_p that minimize $g(m_p, n_p)$ such that

$$g(m_p, n_p) = \big|d_p(f_1) + m_p d_{\max}(f_1) - d_p(f_2) - n_p d_{\max}(f_2)\big|, \qquad (2.14)$$

2.3 Phase Unwrapping from Multiple Depth Maps

Fig. 2.9 Frequency modulation within an integration period. The first half is modulated at f_1, and the other half is modulated at f_2

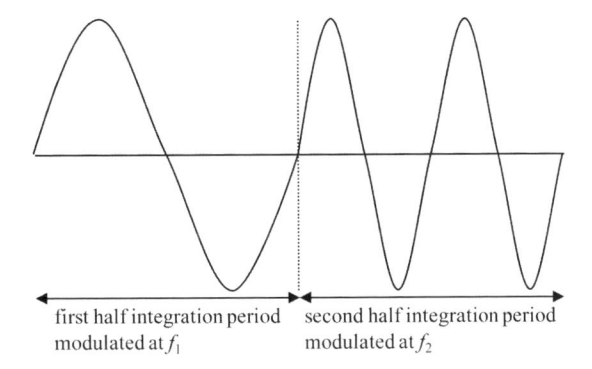

first half integration period modulated at f_1 second half integration period modulated at f_2

where $d_p(f_1)$ and $d_p(f_2)$ denote the measured distances at pixel p in M_1 and M_2 respectively, and $d_{\max}(f)$ denotes the maximum range of f.

The depth consistency constraint has been mentioned by Göktürk et al. [11] and used by Falie and Buzuloiu [8] for phase unwrapping of ToF cameras. The illuminating power of ToF cameras is, however, limited due to the eye-safety problem, and the reflectivity of the scene may be very low. In this situation, the amount of noise may be too large for accurate numbers of wrappings to minimize $g(m_p, n_p)$. For robust estimation against noise, Droeschel et al. [6] incorporate the depth consistency constraint into their earlier work [7] for a single depth map, using an auxiliary depth map of a different modulation frequency.

If we acquire a pair of depth maps of a dynamic scene sequentially and independently, the pixels at the same location may not correspond to each other. To deal with such dynamic situations, several approaches [4, 20] acquire a pair of depth maps simultaneously. These can be divided into single-camera and multicamera methods, as described below.

2.3.1 Single-Camera Methods

For obtaining a pair of depth maps sequentially, four samples of integrated electric charge are required per each integration period, resulting in eight samples within a pair of two different integration periods. Payne et al. [20] propose a special hardware system that enables simultaneous acquisition of a pair of depth maps at different frequencies by dividing the integration period into two, switching between frequencies f_1 and f_2, as shown in Fig. 2.9.

Payne et al. [20] also shows that it is possible to obtain a pair of depth maps with only five or six samples within a combined integration period, using their system. By using fewer samples, the total readout time is reduced and the integration period for each sample can be extended, resulting in an improved signal-to-noise ratio.

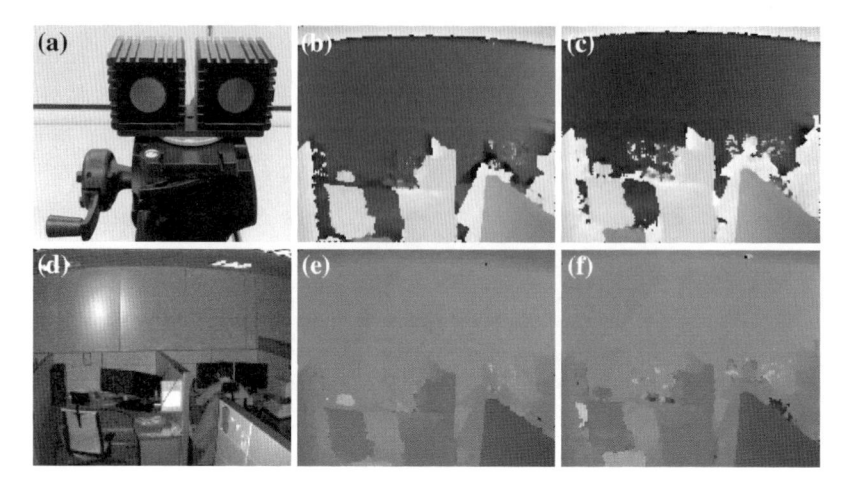

Fig. 2.10 a Stereo ToF camera system. (**b, c**) Depth maps acquired by the system. **d** Amplitude image corresponding to (b). (**e, f**) Unwrapped depth maps, corresponding to (b) and (c), respectively. The intensity in (b, c, e, f) is proportional to the depth. The maximum intensity (255) in (b, c) and (e, f) correspond to 5.2 and 15.6 m, respectively. Images courtesy of Choi and Lee [4]

2.3.2 Multicamera Methods

Choi and Lee [4] use a *pair* of commercially available ToF cameras to simultaneously acquire a pair of depth maps from different viewpoints. The two cameras C_1 and C_2 are fixed to each other, and the mapping of a 3D point $\mathbf{X}$ from C_1 to its corresponding point $\mathbf{X}'$ from C_2 is given by $(\mathbf{R}, \mathbf{T})$, where $\mathbf{R}$ is a 3×3 rotation matrix, and $\mathbf{T}$ is a 3×1 translation vector. In [4], the extrinsic parameters $\mathbf{R}$ and $\mathbf{T}$ are assumed to have been estimated. Figure 2.10a shows the stereo ToF camera system.

Denoting by M_1 and M_2 a pair of depth maps acquired by the system, a pixel p in M_1 and its corresponding pixel q in M_2 should satisfy:

$$\mathbf{X}'_q(n_q) = \mathbf{R}\mathbf{X}_p(m_p) + \mathbf{T}, \tag{2.15}$$

where $\mathbf{X}_p(m_p)$ and $\mathbf{X}'_q(n_q)$ denote the unwrapped 3D points of p and q with their numbers of wrappings m_p and n_q, respectively.

Based on the relation in Eq. (2.15), Choi and Lee [4] generalize the depth consistency constraint in Eq. (2.14) for a single camera to those for the stereo camera system:

$$D_p(m_p) = \min_{n_{q^\star} \in \{0,\ldots,N\}} \left(\left\| \mathbf{X}'_{q^\star}(n_{q^\star}) - \mathbf{R}\mathbf{X}_p(m_p) - \mathbf{T} \right\| \right), \tag{2.16}$$

$$D_q(n_q) = \min_{m_{p^\star} \in \{0,\ldots,N\}} \left(\left\| \mathbf{X}_{p^\star}(m_{p^\star}) - \mathbf{R}^T (\mathbf{X}'_q(n_q) - \mathbf{T}) \right\| \right),$$

2.3 Phase Unwrapping from Multiple Depth Maps

Table 2.1 Summary of phase-unwrapping methods

Methods	# Depth maps	Cues	Approach	Maximum range
Poppinga and Birk [21]	1	CA[a]	Thresholding	$2d_{max}$
Choi et al. [5]	1	CA, DD[b]	Segmentation, MRF	$(N^d + 1)d_{max}$
McClure et al. [17]	1	CA	Segmentation, thresholding	$2d_{max}$
Jutzi [14]	1	DD	Branch cuts, integration	∞
Droeschel et al. [7]	1	DD	MRF, integration	∞
Droeschel et al. [6]	2 (Multi-freq.)	DD, DC[c]	MRF, integration	∞
Payne et al. [20]	2 (Multi-freq.)	DC	Hardware	$\frac{c}{2|f_1-f_2|}$
Choi and Lee [4]	2 (Stereo)	DC	Stereo ToF, MRF	$(N + 1)d_{max}$

[a] Corrected amplitude. [b] Depth discontinuity. [c] Depth consistency. [d] The maximum number of wrappings determined by the user.

where pixels $q^\star$ and $p^\star$ are the projections of $\mathbf{R}\mathbf{X}_p(m_p) + \mathbf{T}$ and $\mathbf{R}^T(\mathbf{X}'_q(n_q) - \mathbf{T})$ onto M_2 and M_1, respectively. The integer N is the maximum number of wrappings, determined by approximate knowledge on the scale of the scene.

To robustly handle with noise and occlusion, Choi and Lee [4] minimize the following MRF energy functions E_1 and E_2, instead of independently minimizing $D_p(m_p)$ and $D_q(m_q)$ at each pixel:

$$E_1 = \sum_{p \in M_1} \hat{D}_p(m_p) + \sum_{(p,u)} V(m_p, m_u), \qquad (2.17)$$

$$E_2 = \sum_{q \in M_2} \hat{D}_q(n_q) + \sum_{(q,v)} V(n_q, n_v),$$

where $\hat{D}_p(m_p)$ and $\hat{D}_q(n_q)$ are the data cost of assigning m_p and n_q to pixels p and q, respectively. Functions $V(m_p, m_u)$ and $V(n_q, n_v)$ determine the discontinuity cost of assigning (m_p, m_u) and (n_q, n_v) to pairs of adjacent pixels (p,u) and (q,v), respectively.

The data costs $\hat{D}_p(m_p)$ and $\hat{D}_q(n_q)$ are defined by truncating $D_p(m_p)$ and $D_q(n_q)$ to prevent their values from becoming too large, due to noise and occlusion:

$$\hat{D}_p(m_p) = \tau_\varepsilon(D_p(m_p)), \quad \hat{D}_q(n_q) = \tau_\varepsilon(D_q(n_q)), \qquad (2.18)$$

$$\tau_\varepsilon(x) = \begin{cases} x, & \text{if } x < \varepsilon, \\ \varepsilon, & \text{otherwise,} \end{cases} \qquad (2.19)$$

where ε is a threshold proportional to the extrinsic calibration error of the system.

The function $V(m_p, m_u)$ is defined in a manner that preserves depth continuity between adjacent pixels. Choi and Lee [4] assume a pair of measured 3D points $\mathbf{X}_p$ and $\mathbf{X}_u$ to have been projected from close surface points if they are close to each other and have similar corrected amplitude values. The proximity is preserved by penalizing the pair of pixels if they have different numbers of wrappings:

$$V(m_p, m_u) = \begin{cases} \frac{\lambda}{r_{pu}} \exp\left(-\frac{\Delta \mathbf{X}_{pu}^2}{2\sigma_{\mathbf{X}}^2}\right) \exp\left(-\frac{\Delta A_{pu}'^2}{2\sigma_{A'}^2}\right) & \text{if } \begin{cases} m_p \neq m_u \text{ and} \\ \Delta \mathbf{X}_{pu} < 0.5\, d_{\max}(f_1) \end{cases} \\ 0 & \text{otherwise.} \end{cases}$$

where λ is a constant, $\Delta \mathbf{X}_{pu}^2 = \|\mathbf{X}_p - \mathbf{X}_u\|^2$, and $\Delta A_{pu}'^2 = \|A_p' - A_u'\|^2$. The variances $\sigma_{\mathbf{X}}^2$ and $\sigma_{A'}^2$ are adaptively determined. The positive scalar r_{pu} is the image coordinate distance between p and u for attenuation of the effect of less adjacent pixels. The function $V(n_q, n_v)$ is defined by analogy with $V(m_p, m_u)$.

Choi and Lee [4] minimize the MRF energies via the α-expansion algorithm [2], obtaining a pair of unwrapped depth maps. To enforce further consistency between the unwrapped depth maps, they iteratively update the MRF energy corresponding to a depth map, using the unwrapped depth of the other map, and perform the minimization until the consistency no longer increases. Figure 2.10e, f shows examples of unwrapped depth maps, as obtained by the iterative optimizations. An alternative method for improving the depth accuracy using two ToF cameras is described in [3].

2.3.3 Discussion

Table 2.1 summarizes the phase-unwrapping methods [4–7, 14, 17, 20, 21] for ToF cameras. The last column of the table shows the extended maximum range, which can be theoretically achieved by the methods. The methods [6, 7, 14] based on the classical phase-unwrapping methods [9, 12] deliver the widest maximum range. In [4, 5], the maximum number of wrappings can be determined by the user. It follows that the maximum range of the methods can also become sufficiently wide, by setting N to a large value. In practice, however, the limited illuminating power of commercially available ToF cameras prevents distant objects from being precisely measured. This means that the phase values may be invalid, even if they can be unwrapped. In addition, the working environment may be physically confined. For the latter reason, Droeschel et al. [6, 7] limit the maximum range to $2d_{\max}$.

2.4 Conclusions

Although the hardware system in [20] has not yet been established in commercially available ToF cameras, we believe that future ToF cameras will use such a frequency modulation technique for accurate and precise depth measurement. In addition, the phase-unwrapping methods in [4, 6] are ready to be applied to a pair of depth maps acquired by such future ToF cameras, for robust estimation of the unwrapped depth values. We believe that a suitable combination of hardware and software systems will extend the maximum ToF range, up to a limit imposed by the illuminating power of the device.

References

1. Bilmes, J.: A gentle tutorial of the EM algorithm and its application to parameter estimation for gaussian mixture and hidden Markov models. Technical Report TR-97-021, University of California. Berkeley (1998)
2. Boykov, Y., Veksler, O., Zabih, R.: Fast approximate energy minimization via graph cuts. IEEE Trans. Pattern Anal. Mach. Intell **23**(11), 1222–1239 (2001)
3. Castañeda, V., Mateus, D., Navab, N.: Stereo time-of-flight. In: Proceedings of the International Conference on Computer Vision (ICCV), pp. 1684–1691 (2011)
4. Choi, O., Lee, S.: Wide range stereo time-of-flight camera. In: Proceedings of International Conference on Image Processing (ICIP) (2012)
5. Choi, O., Lim, H., Kang, B., Kim, Y., Lee, K., Kim, J., Kim, C.: Range unfolding for time-of-flight depth cameras. In: Proceedings of International Conference on Image Processing (ICIP), pp. 4189–4192 (2010)
6. Droeschel, D., Holz, D., Behnke, S.: Multifrequency phase unwrapping for time-of-flight cameras. In: Proceedings of IEEE/RSJ International Conference on Intelligent Robots and Systems (IROS), Taipei (2010)
7. Droeschel, D., Holz, D., Behnke, S.: Probabilistic phase unwrapping for time-of-flight cameras. In: Joint 41st International Symposium on Robotics and 6th German Conference on Robotics (2010)
8. Fălie, D., Buzuloiu, V.: Wide range time of flight camera for outdoor surveillance. In: Microwaves, Radar and Remote Sensing Symposium, pp. 79–82 (2008)
9. Frey, B.J., Koetter, R., Petrovic, N.: Very loopy belief propagation for unwrapping phase images. In: Advances in Neural Information Processing Systems (2001)
10. Ghiglia, D.C., Romero, L.A.: Robust two-dimensional weighted and unweighted phase unwrapping that uses fast transforms and iterative methods. J.Opt. Soc. Am. A **11**(1), 107–117 (1994)
11. Göktürk, S.B., Yalcin, H., Bamji, C.: A time-of-flight depth sensor–system description, issues and solutions. In: Proceedings of the Computer Vision and Parallel Recognition (CVPR) Workshops (2004)
12. Goldstein, R.M., Zebker, H.A., Werner, C.L.: Satellite radar interferometry: two-dimensional phase unwrapping. Radio Sci. **23**, 713–720 (1988)
13. Jakowatz Jr, C., Wahl, D., Eichel, P., Ghiglia, D., Thompson, P.: Spotlight-mode Synthetic Aperture Radar: A Signal Processing Approach. Kluwer Academic Publishers, Boston (1996)
14. Jutzi, B.: Investigation on ambiguity unwrapping of range images. In: International Archives of Photogrammetry and Remote Sensing Workshop on Laserscanning (2009)
15. Liang, V., Lauterbur, P.: Principles of Magnetic Resonance Imaging: A Signal Processing Perspective. Wiley-IEEE Press, New York (1999)
16. Mesa Imaging AG. http://www.mesa-imaging.ch
17. McClure, S.H., Cree, M.J., Dorrington, A.A., Payne, A.D.: Resolving depth-measurement ambiguity with commercially available range imaging cameras. In: Image Processing: Machine Vision Applications III (2010)
18. Meyer, F.: Topographic distance and watershed lines. Signal Process. **38**(1), 113–125 (1994)
19. Oprişescu, Ş., Fălie, D., Ciuc, M., Buzuloiu, V.: Measurements with ToF cameras and their necessary corrections. In: IEEE International Symposium on Signals, Circuits & Systems (2007)
20. Payne, A.D., Jongenelen, A.P.P., Dorrington, A.A., Cree, M.J., Carnegie, D.A.: Multiple frequency range imaging to remove measurment ambiguity. In: 9th Conference on Optical 3-D, Measurement Techniques (2009)
21. Poppinga, J., Birk, A.: A novel approach to efficient error correction for the swissranger time-of-flight 3D camera. In: RoboCup 2008: Robot Soccer World Cup XII (2008)
22. Rother, C., Kolmogorov, V., Blake, A.: "GrabCut"—interactive foreground extraction using iterated graph cuts. In: International Conference and Exhibition on Computer Graphics and Interactive Techniques (2004)

Chapter 3
Calibration of Time-of-Flight Cameras

Abstract This chapter describes the metric calibration of a time-of-flight camera including the internal parameters, and lens distortion. Once the camera has been calibrated, the 2D depth image can be transformed into a range map, which encodes the distance to the scene along each optical ray. It is convenient to use established calibration methods, which are based on images of a chequerboard pattern. The low resolution of the amplitude image, however, makes it difficult to detect the board reliably. Heuristic detection methods, based on connected image components, perform very poorly on this data. An alternative, geometrically principled method is introduced here, based on the Hough transform. The Hough method is compared to the standard OpenCV board-detection routine, by application to several hundred time-of-flight images. It is shown that the new method detects significantly more calibration boards, over a greater variety of poses, without any significant loss of accuracy.

Keywords Time-of-Flight cameras · Low-resolution camera calibration · Hough transform · Chessboard detection

3.1 Introduction

Time-of-Flight (ToF) cameras can, in principle, be modeled and calibrated as pinhole devices. For example, if a known chequerboard pattern is detected in a sufficient variety of poses, then the internal and external camera parameters can be estimated by standard routines [10, 19]. This chapter will briefly review the underlying calibration model, before addressing the problem of chequerboard detection in detail. The latter is the chief obstacle to the use of existing calibration software, owing to the low resolution of the ToF images.

3.2 Camera Model

If the scene coordinates of a point are $(X, Y, Z)^\top$, then the pinhole projection can be expressed as $(x_p, y_p, 1)^\top \simeq \boldsymbol{R}(X, Y, Z)^\top + \boldsymbol{T}$ where the rotation matrix $\boldsymbol{R}$ and translation $\boldsymbol{T}$ account for the pose of the camera. The observed pixel coordinates of the point are then modeled as

$$\begin{pmatrix} x \\ y \\ 1 \end{pmatrix} = \begin{pmatrix} f s_x & f s_\theta & x_0 \\ 0 & f s_y & y_0 \\ 0 & 0 & 1 \end{pmatrix} \begin{pmatrix} x_d \\ y_d \\ 1 \end{pmatrix} \qquad (3.1)$$

where $(x_d, y_d)^\top$ results from lens distortion of $(x_p, y_p)^\top$. The parameter f is the focal length, (s_x, s_y) are the pixel scales, and s_θ is the skew factor [10], which is assumed to be zero here. The lens distortion may be modeled by a radial part $\boldsymbol{d}_1$ and tangential part $\boldsymbol{d}_2$, so that

$$\begin{pmatrix} x_d \\ y_d \end{pmatrix} = d_1(r) \begin{pmatrix} x_p \\ y_p \end{pmatrix} + \boldsymbol{d}_2(x_p, y_p) \quad \text{where} \quad r = \sqrt{x_p^2 + y_p^2} \qquad (3.2)$$

is the radial coordinate. The actual distortion functions are polynomials of the form

$$d_1(r) = 1 + a_1 r^2 + a_2 r^4 \quad \text{and} \quad \boldsymbol{d}_2(x, y) = \begin{pmatrix} 2xy & r^2 + 2x^2 \\ r^2 + 2y^2 & 2xy \end{pmatrix} \begin{pmatrix} a_3 \\ a_4 \end{pmatrix}. \qquad (3.3)$$

The coefficients (a_1, a_2, a_3, a_4) must be estimated along with the other internal parameters (f, s_x, s_y) and (x_0, y_0) in (3.1). The standard estimation procedure is based on the projection of a known chequerboard pattern, which is viewed in many different positions and orientations. The external parameters $(\boldsymbol{R}, \boldsymbol{T})$, as well as the internal parameters can then be estimated as described by Zhang [1, 19], for example.

3.3 Board Detection

It is possible to find the chequerboard vertices, in ordinary images, by first detecting image corners [9], and subsequently imposing global constraints on their arrangement [1, 14, 18]. This approach, however, is not reliable for low-resolution images (e.g., in the range 100–500px^2) because the local image structure is disrupted by sampling artifacts, as shown in Fig. 3.1. Furthermore, these artifacts become worse as the board is viewed in distant and slanted positions, which are essential for high-quality calibration [2]. This is a serious obstacle for the application of existing calibration methods to new types of camera. For example, the amplitude signal from a typical ToF camera [15] resembles an ordinary grayscale image, but is of very low spatial resolution (e.g., 176×144), as well as being noisy. It is, nonetheless, necessary to

3.3 Board Detection

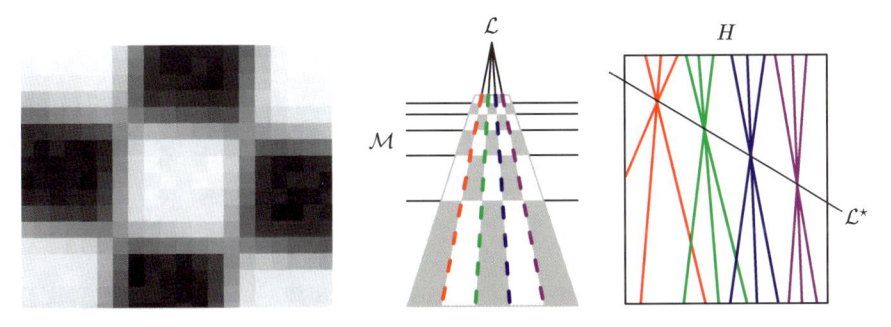

Fig. 3.1 *Left* Example chequers from a TOF amplitude image. Note the variable appearance of the four junctions at this resolution, e.g., '×' at *lower-left* versus '+' at *top-right*. *Middle* A perspective image of a calibration grid is represented by line pencils $\mathscr{L}$ and $\mathscr{M}$, which intersect at the $\ell \times m = 20$ internal vertices of this board. Strong image gradients are detected along the *dashed lines*. *Right* The Hough transform H of the image points associated with $\mathscr{M}$. Each high-gradient point maps to a line, such that there is a pencil in H for each set of edge points. The line $\mathscr{L}^*$, which passes through the $\ell = 4$ Hough vertices, is the Hough representation of the image pencil $\mathscr{L}$

calibrate these devices, in order to combine them with ordinary color cameras, for 3-D modeling and rendering [4, 7, 8, 12, 13, 16, 20].

The method described here is based on the Hough transform [11], and effectively fits a *global* model to the lines in the chequerboard pattern. This process is much less sensitive to the resolution of the data, for two reasons. First, information is integrated across the source image, because each vertex is obtained from the intersection of two fitted lines. Second, the structure of a straight edge is inherently simpler than that of a corner feature. However, for this approach to be viable, it is assumed that any lens distortion has been precalibrated, so that the images of the pattern contain straight lines. This is not a serious restriction, for two reasons. First, it is relatively easy to find enough boards (by any heuristic method) to get adequate estimates of the internal and lens parameters. Indeed, this can be done from a single image, in principle [5]. The harder problems of reconstruction and relative orientation can then be addressed after adding the newly detected boards, ending with a bundle adjustment that also refines the initial internal parameters. Second, the TOF devices used here have fixed lenses, which are sealed inside the camera body. This means that the internal parameters from previous calibrations can be reused.

Another Hough method for chequerboard detection has been presented by de la Escalera and Armingol [3]. Their algorithm involves a *polar* Hough transform of all high-gradient points in the image. This results in an array that contains a peak for each line in the pattern. It is not, however, straightforward to extract these peaks, because their location depends strongly on the unknown orientation of the image lines. Hence, all local maxima are detected by morphological operations, and a second Hough transform is applied to the resulting data in [3]. The true peaks will form two collinear sets in the first transform (cf. Sect. 3.3.5), and so the final task is to detect two peaks in the second Hough transform [17].

The method described in this chapter is quite different. It makes use of the gradient *orientation* as well as magnitude at each point, in order to establish an axis-aligned coordinate system for each image of the pattern. Separate Hough transforms are then performed in the x- and y-directions of the local coordinate system. By construction, the slope coordinate of any line is close to zero in the corresponding *Cartesian* Hough transform. This means that, on average, the peaks occur along a fixed axis of each transform, and can be detected by a simple sweep-line procedure. Furthermore, the known $\ell \times m$ structure of the grid makes it easy to identify the optimal sweep line in each transform. Finally, the two optimal sweep lines map directly back to pencils of ℓ and m lines in the original image, owing to the Cartesian nature of the transform. The principle of the method is shown in Fig. 3.1.

It should be noted that the method presented here was designed specifically for use with TOF cameras. For this reason, the *range*, as well as intensity data is used to help segment the image in Sect. 3.3.2. However, this step could easily be replaced with an appropriate background subtraction procedure [1], in which case the new method could be applied to ordinary RGB images. Camera calibration is typically performed under controlled illumination conditions, and so there would be no need for a dynamic background model.

3.3.1 Overview

The new method is described in Sect. 3.3; preprocessing and segmentation are explained in Sects. 3.3.2 and 3.3.3, respectively, while Sect. 3.3.4 describes the geometric representation of the data. The necessary Hough transforms are defined in Sect. 3.3.5, and analyzed in Sect. 3.3.6.

Matrices and vectors will be written in bold, e.g., $\mathbf{M}, \mathbf{v}$, and the Euclidean length of $\mathbf{v}$ will be written $|\mathbf{v}|$. Equality up to an overall nonzero scaling will be written $\mathbf{v} \simeq \mathbf{u}$. Image points and lines will be represented in homogeneous coordinates [10], with $\mathbf{p} \simeq (x, y, 1)^\top$ and $\mathbf{l} \simeq (\alpha, \beta, \gamma)$, such that $\mathbf{lp} = 0$ if $\mathbf{l}$ passes through $\mathbf{p}$. The intersection point of two lines can be obtained from the cross-product $(\mathbf{l} \times \mathbf{m})^\top$. An assignment from variable a to variable b will be written $b \leftarrow a$. It will be convenient, for consistency with the pseudocode listings, to use the notation $(m : n)$ for the sequence of integers from m to n inclusive. The 'null' symbol $\varnothing$ will be used to denote undefined or unused variables.

The method described here refers to a chequerboard of $(\ell + 1) \times (m + 1)$ squares, with $\ell < m$. It follows that the *internal* vertices of the pattern are imaged as the ℓm intersection points

$$v_{ij} = \mathbf{l}_i \times \mathbf{m}_j \quad \text{where} \quad \mathbf{l}_i \in \mathscr{L} \quad \text{for} \quad i = 1 : \ell \quad \text{and} \quad \mathbf{m}_j \in \mathscr{M} \quad \text{for} \quad j = 1 : m.$$
(3.4)

3.3 Board Detection

The sets $\mathscr{L}$ and $\mathscr{M}$ are *pencils*, meaning that l_i all intersect at a point p, while m_j all intersect at a point q. Note that p and q are the *vanishing points* of the gridlines, which may be at infinity in the images.

It is assumed that the imaging device, such as a ToF camera, provides a range map D_{ij}, containing distances from the optical center, as well as a luminance-like amplitude map A_{ij}. The images D and A are both of size $I \times J$. All images must be undistorted, as described in the Sect. 3.3.

3.3.2 Preprocessing

The amplitude image A is roughly segmented, by discarding all pixels that correspond to very near or far points. This gives a new image B, which typically contains the board, plus the person holding it:

$$B_{ij} \leftarrow A_{ij} \text{ if } d_0 < D_{ij} < d_1, \quad B_{ij} \leftarrow \varnothing \text{ otherwise.} \quad (3.5)$$

The near-limit d_0 is determined by the closest position for which the board remains fully inside the field-of-view of the camera. The far-limit d_1 is typically set to a value just closer than the far wall of the scene. These parameters need only to be set approximately, provided that the interval $d_1 - d_0$ covers the possible positions of the calibration board.

It is useful to perform a morphologic erosion operation at this stage, in order to partially remove the perimeter of the board. In particular, if the physical edge of the board is not white, then it will give rise to irrelevant image gradients. The erosion radius need only be set approximately, assuming that there is a reasonable amount of white space around the chessboard pattern. The gradient of the remaining amplitude image is now computed, using the simple kernel $\Delta = (-1/2, 0, 1/2)$. The horizontal and vertical components are

$$\begin{aligned} \xi_{ij} &\leftarrow (\Delta \star B)_{ij} \\ &= \rho \cos \theta \end{aligned} \quad \text{and} \quad \begin{aligned} \eta_{ij} &\leftarrow (\Delta^\top \star B)_{ij} \\ &= \rho \sin \theta \end{aligned} \quad (3.6)$$

where $\star$ indicates convolution. No presmoothing of the image is performed, owing to the low spatial resolution of the data.

3.3.3 Gradient Clustering

The objective of this section is to assign each gradient vector (ξ_{ij}, η_{ij}) to one of three classes, with labels $\kappa_{ij} \in \{\lambda, \mu, \varnothing\}$. If $\kappa_{ij} = \lambda$ then pixel (i, j) is on one of the lines in

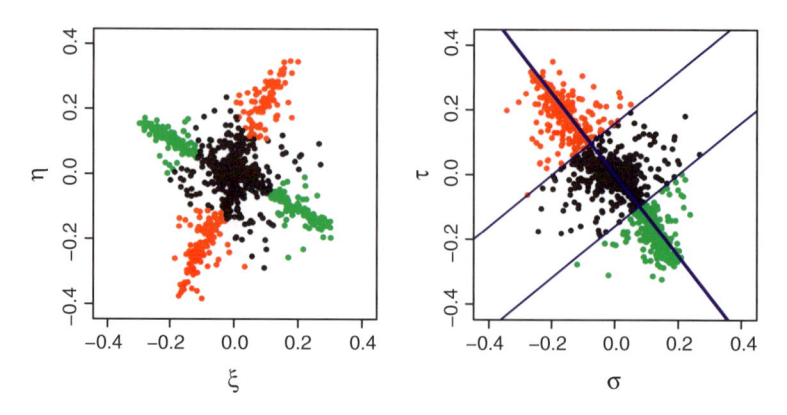

Fig. 3.2 *Left* the cruciform distribution of image gradients, due to *black/white* and *white/black* transitions at each orientation, would be difficult to segment in terms of horizontal and vertical components (ξ, η). *Right* the same distribution is easily segmented, by eigenanalysis, in the double-angle representation (3.7). The *red* and *green* labels are applied to the corresponding points in the original distribution, on the *left*

$\mathscr{L}$, and (ξ_{ij}, η_{ij}) is perpendicular to that line. If $\kappa_{ij} = \mu$, then the analogous relations hold with respect to $\mathscr{M}$. If $\kappa_{ij} = \varnothing$ then pixel (i, j) does not lie on any of the lines.

The gradient distribution, after the initial segmentation, will contain two elongated clusters through the origin, which will be approximately orthogonal. Each *cluster* corresponds to a gradient orientation (mod π), while each *end* of a cluster corresponds to a gradient polarity (black/white vs. white/black). The distribution is best analyzed after a double-angle mapping [6], which will be expressed as $(\xi, \eta) \mapsto (\sigma, \tau)$. This mapping results in a *single* elongated cluster, each *end* of which corresponds to a gradient orientation (mod π), as shown in Fig. 3.2. The double-angle coordinates are obtained by applying the trigonometric identities $\cos(2\theta) = \cos^2\theta - \sin^2\theta$ and $\sin(2\theta) = 2\sin\theta\cos\theta$ to the gradients (3.6), so that

$$\sigma_{ij} \leftarrow \frac{1}{\rho_{ij}}(\xi_{ij}^2 - \eta_{ij}^2) \quad \text{and} \quad \tau_{ij} \leftarrow \frac{2}{\rho_{ij}}\xi_{ij}\eta_{ij} \quad \text{where} \quad \rho_{ij} = \sqrt{\xi_{ij}^2 + \eta_{ij}^2} \qquad (3.7)$$

for all points at which the magnitude ρ_{ij} is above machine precision. Let the first unit eigenvector of the (σ, τ) covariance matrix be $(\cos(2\phi), \sin(2\phi))$, which is written in this way so that the angle ϕ can be interpreted in the original image. The cluster membership is now defined by the projection

$$\pi_{ij} = (\sigma_{ij}, \tau_{ij}) \cdot (\cos(2\phi), \sin(2\phi)) \qquad (3.8)$$

of the data onto this axis. The gradient vectors (ξ_{ij}, η_{ij}) that project to either end of the axis are labeled as follows:

3.3 Board Detection

$$\kappa_{ij} \leftarrow \begin{cases} \lambda & \text{if } \pi_{ij} \geq \rho_{\min} \\ \mu & \text{if } \pi_{ij} \leq -\rho_{\min} \\ \emptyset & \text{otherwise.} \end{cases} \quad (3.9)$$

Strong gradients that are not aligned with either axis of the board are assigned to $\emptyset$, as are all weak gradients. It should be noted that the respective *identity* of classes λ and μ has not yet been determined; the correspondence $\{\lambda, \mu\} \Leftrightarrow \{\mathscr{L}, \mathscr{M}\}$ between labels and pencils will be resolved in Sect. 3.3.6.

3.3.4 Local Coordinates

A coordinate system will now be constructed for each image of the board. The very low amplitudes $B_{ij} \approx 0$ of the *black* squares tend to be characteristic of the board (i.e., $B_{ij} \gg 0$ for both the white squares and for the rest of B). Hence, a good estimate of the center can be obtained by normalizing the amplitude image to the range $[0, 1]$ and then computing a centroid using weights $(1 - B_{ij})$. The centroid, together with the angle ϕ from (3.8) defines the Euclidean transformation $(x, y, 1)^\top = \boldsymbol{E}\,(j, i, 1)^\top$ into local coordinates, centered on and aligned with the board.

Let $(x_\kappa, y_\kappa, 1)^\top$ be the coordinates of point (i, j), after transformation by $\boldsymbol{E}$, with the label κ inherited from κ_{ij}, and let $\mathscr{L}'$ and $\mathscr{M}'$ correspond to $\mathscr{L}$ and $\mathscr{M}$ in the new coordinate system. Now, by construction, any labeled point is hypothesized to be part of $\mathscr{L}'$ or $\mathscr{M}'$, such that $\boldsymbol{l}'(x_\lambda, y_\lambda, 1)^\top = 0$ or $\boldsymbol{m}'(x_\mu, y_\mu, 1)^\top = 0$, where $\boldsymbol{l}'$ and $\boldsymbol{m}'$ are the local coordinates of the relevant lines $\boldsymbol{l}$ and $\boldsymbol{m}$, respectively. These lines can be expressed as

$$\boldsymbol{l}' \simeq (-1, \beta_\lambda, \alpha_\lambda) \quad \text{and} \quad \boldsymbol{m}' \simeq (\beta_\mu, -1, \alpha_\mu) \quad (3.10)$$

with inhomogeneous forms $x_\lambda = \alpha_\lambda + \beta_\lambda y_\lambda$ and $y_\mu = \alpha_\mu + \beta_\mu x_\mu$, such that the slopes $|\beta_\kappa| \ll 1$ are *bounded*. In other words, the board is axis aligned in local coordinates, and the perspective-induced deviation of any line is less than $45°$.

3.3.5 Hough Transform

The Hough transform, in the form used here, maps *points* from the image to *lines* in the transform. In particular, points along a line are mapped to lines through a point. This duality between collinearity and concurrency suggests that a *pencil* of n image lines will be mapped to a *line* of n transform points, as in Fig. 3.1.

The transform is implemented as a 2-D histogram $H(u, v)$, with horizontal and vertical coordinates $u \in [0, u_1]$ and $v \in [0, v_1]$. The point $(u_0, v_0) = \frac{1}{2}(u_1, v_1)$ is the center of the transform array. Two transforms, H_λ and H_μ, will be performed, for

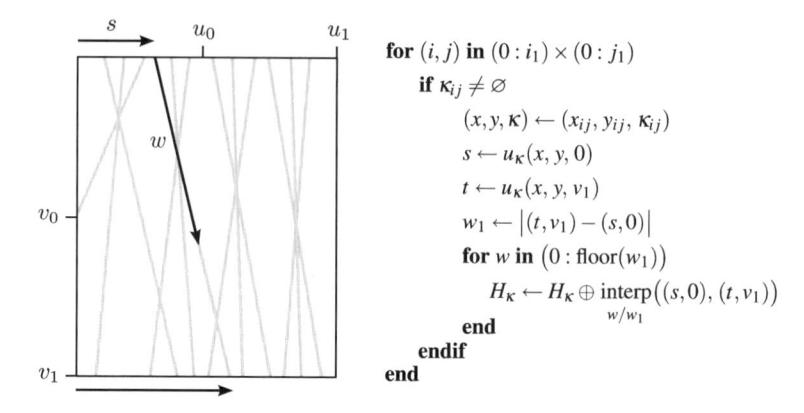

```
    s      u₀         u₁             for (i, j) in (0 : i₁) × (0 : j₁)
                                        if κᵢⱼ ≠ ∅
                                           (x, y, κ) ← (xᵢⱼ, yᵢⱼ, κᵢⱼ)
         w                                 s ← u_κ(x, y, 0)
                                           t ← u_κ(x, y, v₁)
  v₀                                       w₁ ← |(t, v₁) − (s, 0)|
                                           for w in (0 : floor(w₁))
                                              H_κ ← H_κ ⊕ interp((s,0), (t,v₁))
                                                                w/w₁
                                           end
                                        endif
  v₁                                  end
```

Fig. 3.3 Hough transform. Each gradient pixel (x, y) labeled $\kappa \in \{\lambda, \mu\}$ maps to a line $u_\kappa(x, y, v)$ in transform H_κ. The operators $H \oplus p$ and $\text{interp}_\alpha(p, q)$ perform accumulation and linear interpolation, respectively. See Sect. 3.3.5 for details

points labeled λ and μ, respectively. The Hough variables are related to the image coordinates in the following way:

$$u_\kappa(x, y, v) = \begin{cases} u(x, y, v) & \text{if } \kappa = \lambda \\ u(y, x, v) & \text{if } \kappa = \mu \end{cases} \quad \text{where} \quad u(x, y, v) = u_0 + x - y(v - v_0). \tag{3.11}$$

Here, $u(x, y, v)$ is the u-coordinate of a line (parameterized by v), which is the Hough transform of an image point (x, y). The Hough intersection point $(u_\kappa^\star, v_\kappa^\star)$ is found by taking two points (x, y) and (x', y'), and solving $u_\lambda(x, y, v) = u_\lambda(x', y', v)$, with x_λ and x'_λ substituted according to (3.10). The same coordinates are obtained by solving $u_\mu(x, y, v) = u_\mu(x', y', v)$, and so the result can be expressed as

$$u_\kappa^\star = u_0 + \alpha_\kappa \quad \text{and} \quad v_\kappa^\star = v_0 + \beta_\kappa \tag{3.12}$$

with labels $\kappa \in \{\lambda, \mu\}$ as usual. A peak at $(u_\kappa^\star, v_\kappa^\star)$ evidently maps to a line of intercept $u_\kappa^\star - u_0$ and slope $v_\kappa^\star - v_0$. Note that if the perspective distortion in the images is small, then $\beta_\kappa \approx 0$, and all intersection points lie along the horizontal midline (u, v_0) of the corresponding transform. The Hough intersection point $(u_\kappa^\star, v_\kappa^\star)$ can be used to construct an image line l' or m', by combining (3.12) with (3.10), resulting in

$$l' \leftarrow \left(-1, v_\lambda^\star - v_0, u_\lambda^\star - u_0\right) \quad \text{and} \quad m' \leftarrow \left(v_\mu^\star - v_0, -1, u_\mu^\star - u_0\right). \tag{3.13}$$

The transformation of these line vectors, back to the original image coordinates, is given by the inverse transpose of the matrix E, described in Sect. 3.3.4.

The two Hough transforms are computed by the procedure in Fig. 3.3. Let H_κ refer to H_λ or H_μ, according to the label κ of the ij-th point (x, y). For each accepted

3.3 Board Detection

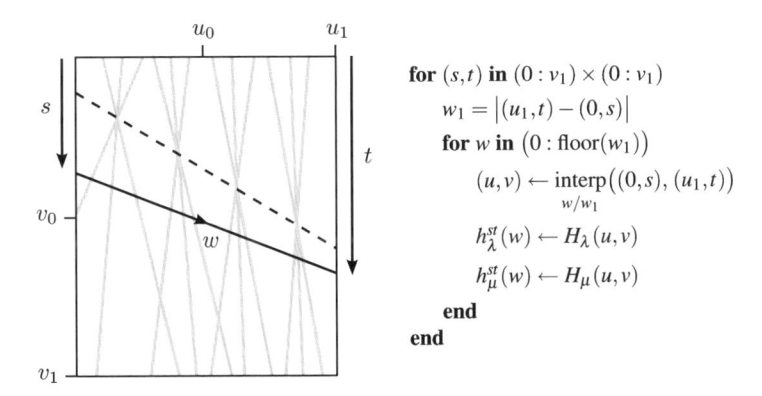

Fig. 3.4 A line $h_\kappa^{st}(w)$, with endpoints $(0, s)$ and (u_1, t), is swept through each Hough transform H_κ. A total of $v_1 \times v_1$ 1-D histograms $h_\kappa^{st}(w)$ are computed in this way. See Sect. 3.3.6 for details

```
for (s,t) in (0 : v₁) × (0 : v₁)
    w₁ = |(u₁,t) − (0,s)|
    for w in (0 : floor(w₁))
        (u,v) ← interp_{w/w₁}((0,s), (u₁,t))
        h_λ^{st}(w) ← H_λ(u,v)
        h_μ^{st}(w) ← H_μ(u,v)
    end
end
```

point, the corresponding line (3.11) intersects the top and bottom of the (u, v) array at points $(s, 0)$ and (t, v_1), respectively. The resulting segment, of length w_1, is evenly sampled, and H_κ is incremented at each of the constituent points. The procedure in Fig. 3.3 makes use of the following functions. First, $\text{interp}_\alpha(p, q)$, with $\alpha \in [0, 1]$, returns the affine combination $(1-\alpha)p + \alpha q$. Second, the 'accumulation' $H \oplus (u, v)$ is equal to $H(u, v) \leftarrow H(u, v) + 1$ if u and v are integers. In the general case, however, the four pixels closest to (u, v) are updated by the corresponding bilinear-interpolation weights (which sum to one).

3.3.6 Hough Analysis

The local coordinates defined in Sect. 3.3.4 ensure that the two Hough transforms H_λ and H_μ have the same characteristic structure. Hence, the subscripts λ and μ will be suppressed for the moment. Recall that each Hough cluster corresponds to a line in the image space, and that a collinear set of Hough clusters corresponds to a pencil of lines in the image space, as in Fig. 3.1. It follows that all lines in a pencil can be detected simultaneously, by *sweeping* the Hough space H with a line that cuts a 1-D slice through the histogram.

Recall from Sect. 3.3.5 that the Hough peaks are most likely to lie along a horizontal axis (corresponding to a fronto-parallel pose of the board). Hence, a suitable parameterization of the sweep line is to vary one endpoint $(0, s)$ along the left edge, while varying the other endpoint (u_1, t) along the right edge, as in Fig. 3.4. This scheme has the desirable property of sampling more densely around the midline (u, v_0). It is also useful to note that the sweep-line parameters s and t can be used to represent the apex of the corresponding pencil. The local coordinates p' and q' are $p' \simeq (l'_s \times l'_t)^\top$ and $q' \simeq (m'_s \times m'_t)^\top$ where l'_s and l'_t are obtained from (3.10) by setting $(u_\lambda^\star, v_\lambda^\star)$ to $(0, s)$ and (u_1, t) respectively, and similarly for m'_s and m'_t.

The procedure shown in Fig. 3.4 is used to analyze the Hough transform. The sweep line with parameters s and t has the form of a 1-D histogram $h_\kappa^{st}(w)$. The integer index $w \in (0 : w_1)$ is equal to the Euclidean distance $|(u, v) - (0, s)|$ along the sweep line. The procedure shown in Fig. 3.4 makes further use of the interpolation operator that was defined in Sect. 3.3.5. Each sweep line $h_\kappa^{st}(w)$, constructed by the above process, will contain a number of isolated clusters: $\text{count}(h_\kappa^{st}) \geq 1$. The clusters are simply defined as runs of nonzero values in $h_\kappa^{st}(w)$. The existence of separating zeros is, in practice, highly reliable when the sweep line is close to the true solution. This is simply because the Hough data was thresholded in (3.9), and strong gradients are not found *inside* the chessboard squares. The representation of the clusters, and subsequent evaluation of each sweep line, will now be described.

The label κ and endpoint parameters s and t will be suppressed, in the following analysis of a single sweep line, for clarity. Hence, let $w \in (a_c : b_c)$ be the interval that contains the c-th cluster in $h(w)$. The score and location of this cluster are defined as the mean value and centroid, respectively:

$$\text{score}(h) = \frac{\sum_{w=a_c}^{b_c} h(w)}{1 + b_c - a_c} \quad \text{and} \quad w_c = a_c + \frac{\sum_{w=a_c}^{b_c} h(w)w}{\sum_{w=a_c}^{b_c} h(w)} \quad (3.14)$$

More sophisticated definitions are possible, based on quadratic interpolation around each peak. However, the mean and centroid give similar results in practice. A total score must now be assigned to the sweep line, based on the scores of the constituent clusters. If n peaks are sought, then the total score is the sum of the highest n cluster scores. But if there are fewer than n clusters in $h(w)$, then this cannot be a solution, and the score is zero:

$$\Sigma^n(h) = \begin{cases} \sum_{i=1}^{n} \text{score}(h) & \text{if } n \leq \text{count}(h) \\ 0 & \text{otherwise} \end{cases} \quad (3.15)$$

where $c(i)$ is the index of the i-th highest-scoring cluster. The optimal clusters are those in the sweep line that maximizes (3.15). Now, restoring the full notation, the score of the optimal sweep line in the transform H_κ is

$$\Sigma_\kappa^n \leftarrow \max_{s,t} \text{score}(h_\kappa^{st}). \quad (3.16)$$

One problem remains: it is not known in advance whether there should be ℓ peaks in H_λ and m in H_μ, or vice versa. Hence all four combinations, $\Sigma_\lambda^\ell, \Sigma_\mu^m, \Sigma_\mu^\ell, \Sigma_\lambda^m$ are computed. The ambiguity between pencils ($\mathscr{L}, \mathscr{M}$) and labels (λ, μ) can then be resolved, by picking the solution with the highest *total* score:

$$(\mathscr{L}, \mathscr{M}) \Leftrightarrow \begin{cases} (\lambda, \mu) & \text{if } \Sigma_\lambda^\ell + \Sigma_\mu^m > \Sigma_\mu^\ell + \Sigma_\lambda^m \\ (\mu, \lambda) & \text{otherwise.} \end{cases} \quad (3.17)$$

3.3 Board Detection

Here, for example, $(\mathscr{L}, \mathscr{M}) \Leftrightarrow (\lambda, \mu)$ means that there is a pencil of ℓ lines in H_λ and a pencil of m lines in H_μ. The procedure in (3.17) is based on the fact that the complete solution must consist of $\ell + m$ clusters. Suppose, for example, that there are ℓ good clusters in H_λ, and m good clusters in H_μ. Of course, there are also ℓ good clusters in H_μ, because $\ell < m$ by definition. However, if only ℓ clusters are taken from H_μ, then an additional $m - \ell$ weak or nonexistent clusters must be found in H_λ, and so the total score $\Sigma_\mu^\ell + \Sigma_\lambda^m$ would not be maximal.

It is straightforward, for each centroid w_c in the optimal sweep line h_κ^{st}, to compute the 2-D Hough coordinates

$$\left(u_\kappa^\star, v_\kappa^\star\right) \leftarrow \underset{w_c/w_1}{\operatorname{interp}}\bigl((0, s), (u_1, t)\bigr) \quad (3.18)$$

where w_1 is the length of the sweep line, as in Fig. 3.4. Each of the resulting ℓm points are mapped to image lines, according to (3.13). The vertices v_{ij} are then be computed from (3.4). The order of intersections along each line is preserved by the Hough transform, and so the ij indexing is automatically consistent.

The final decision function is based on the observation that cross-ratios of distances between consecutive vertices should be near unity (because the images are projectively related to a regular grid). In practice, it suffices to consider simple ratios, taken along the first and last edge of each pencil. If all ratios are below a given threshold, then the estimate is accepted. This threshold was fixed once and for all, such that no false-positive detections (which are unacceptable for calibration purposes) were made, across *all* data sets.

3.3.7 Example Results

The method was tested on five multicamera data sets, and compared to the standard OpenCV detector. Both the OpenCV and Hough detections were refined by the OpenCV subpixel routine, which adjusts the given point to minimize the discrepancy with the image gradient around the chequerboard corner [1, 2]. Table 3.1 shows the number of true-positive detections by each method, as well as the number of detections common to both methods. The geometric error is the discrepancy from the 'ideal' board, after fitting the latter by the optimal (DLT+LM) homography [10]. This is by far the most useful measure, as it is directly related to the role of the detected vertices in subsequent calibration algorithms (and also has a simple interpretation in pixel units). The photometric error is the gradient residual, as described in Sect. 3.3.6. This measure is worth considering, because it is the criterion minimized by the subpixel optimization, but it is less interesting than the geometric error.

The Hough method detects 35 % more boards than the OpenCV method, on average. There is also a slight reduction in average geometric error, even though the additional boards were more problematic to detect. The results should not be surprising, because the new method uses a very strong model of the global board geometry (in fairness, it also benefits from the depth thresholding in 3.3.2). There were zero

Table 3.1 Results over six multi-ToF camera setups

Set / Camera	Number detected			Geometric error		Photometric error	
	OCV	HT	Both	OCV	HT	OCV	HT
1 / 1	19	34	13	0.2263	0.1506	0.0610	0.0782
1 / 2	22	34	14	0.1819	0.1448	0.0294	0.0360
1 / 3	46	33	20	0.1016	0.0968	0.0578	0.0695
1 / 4	26	42	20	0.2044	0.1593	0.0583	0.0705
2 / 1	15	27	09	0.0681	0.0800	0.0422	0.0372
2 / 2	26	21	16	0.0939	0.0979	0.0579	0.0523
2 / 3	25	37	20	0.0874	0.0882	0.0271	0.0254
3 / 1	14	26	11	0.1003	0.0983	0.0525	0.0956
3 / 2	10	38	10	0.0832	0.1011	0.0952	0.1057
3 / 3	25	41	21	0.1345	0.1366	0.0569	0.0454
3 / 4	18	23	10	0.1071	0.1053	0.0532	0.0656
4 / 1	16	21	14	0.0841	0.0874	0.0458	0.0526
4 / 2	45	53	29	0.0748	0.0750	0.0729	0.0743
4 / 3	26	42	15	0.0954	0.0988	0.0528	0.0918
5 / 1	25	37	18	0.0903	0.0876	0.0391	0.0567
5 / 2	20	20	08	0.2125	0.1666	0.0472	0.0759
5 / 3	39	36	24	0.0699	0.0771	0.0713	0.0785
5 / 4	34	35	19	0.1057	0.1015	0.0519	0.0528
6 / 1	29	36	20	0.1130	0.1203	0.0421	0.0472
6 / 2	35	60	26	0.0798	0.0803	0.0785	0.1067
Mean:	25.75	34.8	16.85	0.1157	0.1077	0.0547	0.0659

Total detections for the OpenCV (515) versus Hough Transform (696) method are shown, as well as the accuracy of the estimates. Geometric error is in pixels. The chief conclusion is that the HT method detects 35 % more boards, and slightly reduces the average geometric error

false-positive detections (100 % precision), as explained in Sect. 3.3.6. The number of true negatives is not useful here, because it depends largely on the configuration of the cameras (i.e., how many images show the back of the board). The false negatives do not provide a very useful measure either, because they depend on an arbitrary judgement about which of the very foreshortened boards 'ought' to have been detected (i.e., whether an edge-on board is 'in' the image or not). Some example detections are shown in Figs. 3.5–3.7, including some difficult cases.

3.4 Conclusions

A new method for the automatic detection of calibration grids in ToF images has been described. The method is based on careful reasoning about the global geometric structure of the board, before and after perspective projection. The method detects many more boards than existing heuristic approaches, which results in a larger and more complete data set for subsequent calibration algorithms. Future work will investigate the possibility of making a global refinement of the pencils, in the geometric parameterization, by minimizing a photometric cost function.

3.4 Conclusions

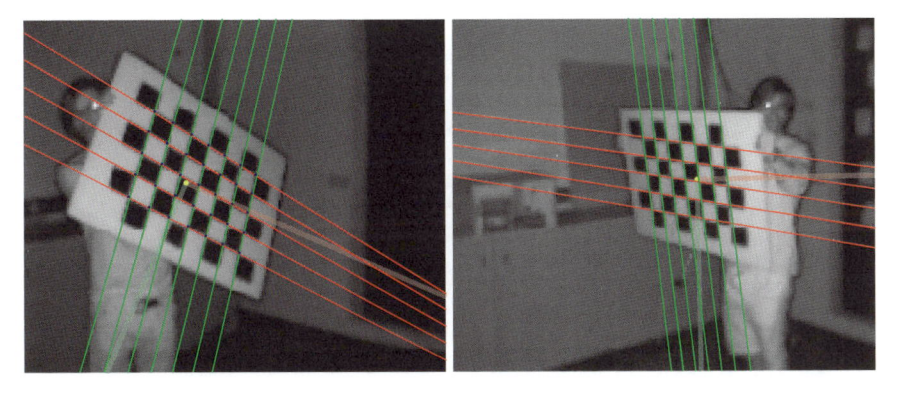

Fig. 3.5 Example detections in 176 × 144 ToF amplitude images. The *yellow dot* (1-pixel radius) is the estimated centroid of the board, and the attached *thick* translucent lines are the estimated axes. The board on the *right*, which is relatively distant and slanted, was not detected by OpenCV

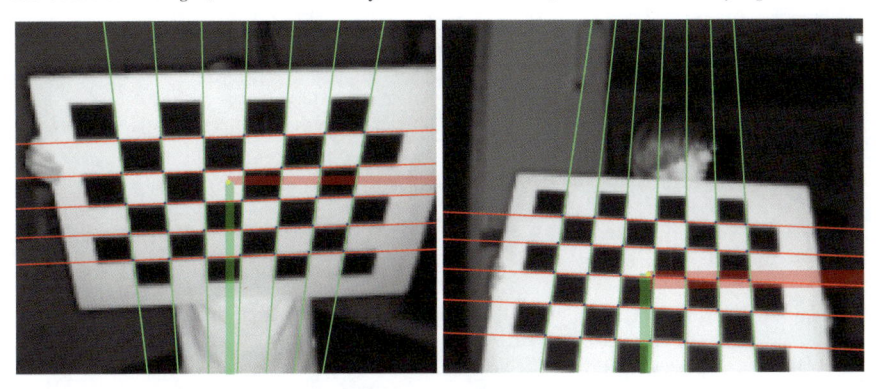

Fig. 3.6 Example detections (cf. Fig. 3.5) showing significant perspective effects

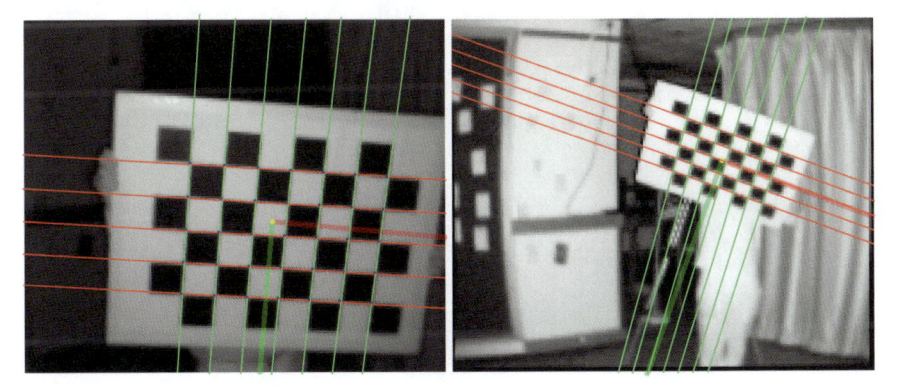

Fig. 3.7 Example detections (cf. Fig. 3.5) showing significant scale changes. The board on the *right*, which is in an image that shows background clutter and lens distortion, was not detected by OpenCV

References

1. Bradski, G., Kaehler, A.: Learning OpenCV. O'Reilly, Sebastopol (2008)
2. Datta, A., Jun-Sik, K., Kanade, T.: Accurate camera calibration using iterative refinement of control points. In: Workshop on Visual Surveillance, Proceedings of IEEE International Conference on Computer Vision (ICCV), pp. 1201–1208 (2009)
3. de la Escalera, A., Armingol, J.: Automatic chessboard detection for intrinsic and extrinsic camera parameter calibration. Sensors **10**, 2027–2044 (2010)
4. Dubois, J.M., Hügli, H.: Fusion of time-of-flight camera point clouds. In: European Conference on Computer Vision (ECCV) Workshop on Multi-Camera and Multi-modal Sensor Fusion Algorithms and Applications (2008)
5. Gonzalez-Aguilera, D., Gomez-Lahoz, J., Rodriguez-Gonzalvez, P.: An automatic approach for radial lens distortion correction from a single image. IEEE Sens. **11**(4), 956–965 (2011)
6. Granlund, G.: In search of a general picture processing operator. Comput. Graph. Image Process. **8**, 155–173 (1978)
7. Hahne, U., Alexa, M.: Depth imaging by combining time-of-flight and on-demand stereo. In: Proceedings of DAGM Workshop on Dynamic 3D Imaging, pp. 70–83 (2009)
8. Hansard, M., Horaud, R., Amat, M., Lee, S.: Projective alignment of range and parallax Data. In: Proceedings of Computer Vision and Parallel Recognition (CVPR), pp. 3089–3096 (2011)
9. Harris, C., Stephens, M.: A combined corner and edge detector. In: Proceedings of the 4th Alvey Vision Conference, pp. 147–151 (1988)
10. Hartley, R., Zisserman, A.: Multiple View Geometry in Computer Vision. Cambridge University Press, Cambridge (2000)
11. Illingworth, J., Kittler, J.: A survey of the hough transform. Comput. Vis. Graph. Image Process. **44**, 87–116 (1988)
12. Koch, R., Schiller, I., Bartczak, B., Kellner, F., Köser, K.: MixIn3D: 3D mixed reality with ToF-camera. In: Proceedings of the DAGM Workshop on Dynamic 3D Imaging, pp. 126–141 (2009)
13. Kolb, A., Barth, E., Koch, R., Larsen, R.: Time-of-flight cameras in computer graphics. Comput. Graph. Forum **29**(1), 141–159 (2010)
14. Krüger, L., Wöhler, C., Würz-Wessel, A., Stein, F.: In-factory calibration of multiocular camera systems. In: Proceedings of the SPIE Photonics Europe, pp. 126–137 (2004)
15. Mesa Imaging AG. http://www.mesa-imaging.ch
16. Schiller, I., Beder, C., Koch, R.: Calibration of a PMD camera using a planar calibration object together with a multi-camera setup. In: International Archieves of Photogrammetry, Remote Sensing and Spatial Information Sciences, XXI, pp. 297–302 (2008)
17. Tuytelaars, T., Proesmans, M., Gool, L.V.: The cascaded hough transform as support for grouping and finding vanishing points and lines. In: Proceedings of International Workshop on Algebraic Frames for the Perception-Action Cycle, pp. 278–289 (1997)
18. Wang, Z., Wu, W., Xu, X., Xue, D.: Recognition and location of the internal corners of planar checkerboard calibration pattern image. Appl. Math. Comput. **185**(2), 894–906 (2007)
19. Zhang, Z.: A flexible new technique for camera calibration. IEEE Trans. Pattern Anal. Mach. Intell. **22**(11), 1330–1334 (2000)
20. Zhu, J., Wang, L., Yang, R.G., Davis, J.: Fusion of time-of-flight depth and stereo for high accuracy depth maps. In: Proceedings of IEEE Conference on Computer Vision and Parallal Recognition (CVPR), pp. 1–8 (2008)

Chapter 4
Alignment of Time-of-Flight and Stereoscopic Data

Abstract An approximately Euclidean representation of the visible scene can be obtained directly from a time-of-flight camera. An uncalibrated binocular system, in contrast, gives only a projective reconstruction of the scene. This chapter analyzes the geometric mapping between the two representations, without requiring an intermediate calibration of the binocular system. The mapping can be found by either of two new methods, one of which requires point correspondences between the range and color cameras, and one of which does not. It is shown that these methods can be used to reproject the range data into the binocular images, which makes it possible to associate high-resolution color and texture with each point in the Euclidean representation. The extension of these methods to multiple time-of-flight system is demonstrated, and the associated problems are examined. An evaluation metric, which distinguishes calibration error from combined calibration and depth error, is developed. This metric is used to evaluate a system that is based on three time-of-flight cameras.

Keywords Depth and color combination · Projective alignment · Time-of-Flight camera calibration · Multicamera systems

4.1 Introduction

It was shown in the preceding chapter that time-of-flight (ToF) cameras can be geometrically calibrated by standard methods. This means that each pixel records an estimate of the scene distance (range) along the corresponding ray, according to the principles described in Chap. 1. The 3-D structure of a scene can also be reconstructed from two or more ordinary images, via the *parallax* between corresponding image points. There are many advantages to be gained by combining the range and parallax data. Most obviously, each point in a parallax-based reconstruction can be mapped back into the original images, from which color and texture can be obtained.

Fig. 4.1 The central panel shows a range image, color-coded according to depth (the *blue* region is beyond the far limit of the device). The *left* and *right* cameras were aligned to the ToF system, using the methods described here. Each 3-D range pixel is reprojected into the high-resolution left and right images (untinted regions were occluded, or otherwise missing, from the range images). Note the large difference between the binocular views, which would be problematic for dense stereo-matching algorithms. It can also be seen that the ToF information is noisy, and of low resolution

Parallax-based reconstructions are, however, difficult to obtain, owing to the difficulty of putting the image points into correspondence. Indeed, it may be impossible to find any correspondences in untextured regions. Furthermore, if a Euclidean reconstruction is required, then the cameras must be calibrated. The accuracy of the resulting reconstruction will also tend to decrease with the distance of the scene from the cameras [23].

The range data, on the other hand, are often very noisy (and, for very scattering surfaces, incomplete), as described in Chap. 1. The spatial resolution of current ToF sensors is relatively low, the depth range is limited, and the luminance signal may be unusable for rendering. It should also be recalled that ToF cameras of the type used here [19] cannot be used in outdoor lighting conditions. These considerations lead to the idea of a *mixed* color and ToF system [18] as shown in Figs. 4.1 and 4.2. Such a system could, in principle, be used to make high-resolution Euclidean reconstructions, with full photometric information [17]. The task of camera calibration would be simplified by the ToF camera, while the visual quality of the reconstruction would be ensured by the color cameras.

In order to make full use of a mixed range/parallax system, it is necessary to find the exact geometric relationship between the different devices. In particular, the reprojection of the ToF data, into the color images, must be obtained. This chapter is concerned with the estimation of these geometric relationships. Specifically, the aim is to align the range and parallax reconstructions, by a suitable 3-D transformation. The alignment problem has been addressed previously, by fully calibrating the binocular system, and then aligning the two reconstructions by a rigid transformation [6, 12, 27, 28]. This approach can be extended in two ways. First, it is possible to optimize over an explicit parameterization of the camera matrices, as in the work of Beder et al. [3] and Koch et al. [16]. The relative position and orientation of all cameras can be estimated by this method. Second, it is possible to minimize an intensity cost between the images and the luminance signal of the ToF camera. This method estimates the photometric, as well as geometric, relationships between the different

4.1 Introduction

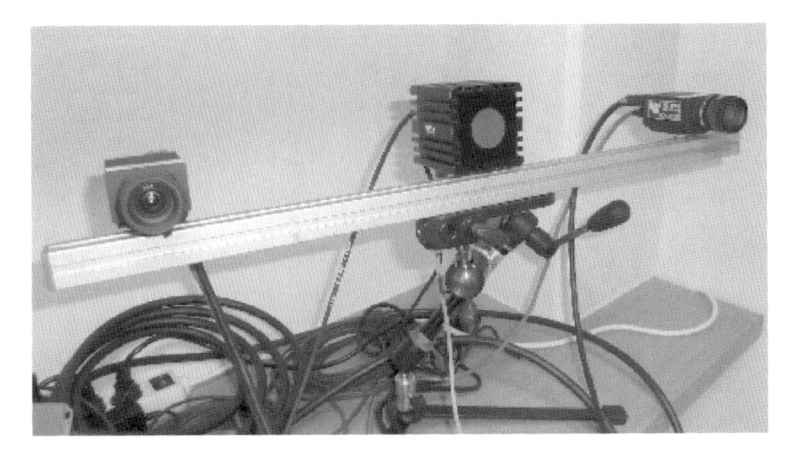

Fig. 4.2 A single ToF+2RGB system, as used in this chapter, with the ToF camera in the center of the rail

cameras [13, 22, 25]. A complete calibration method, which incorporates all of these considerations, is described by Lindner et al. [18].

The approaches described above, while capable of producing good results, have some limitations. First, there may be residual distortions in the range data, that make a rigid alignment impossible [15]. Second, these approaches require the binocular system to be fully calibrated, and recalibrated after any movement of the cameras. This requires, for best results, many views of a known calibration object. Typical view-synthesis applications, in contrast, require only a weak calibration of the cameras. One way to remove the calibration requirement is to perform an essentially 2-D registration of the different images [1, 4]. This, however, can only provide an instantaneous solution, because changes in the scene structure produce corresponding changes in the image-to-image mapping.

An alternative approach is proposed here. It is hypothesized that the ToF reconstruction is approximately Euclidean. This means that an *uncalibrated* binocular reconstruction can be mapped directly into the Euclidean frame, by a suitable 3-D projective transformation. This is a great advantage for many applications, because automatic uncalibrated reconstruction is relatively easy. Furthermore, although the projective model is much more general than the rigid model, it preserves many important relationships between the images and the scene (e.g., epipolar geometry and incidence of points on planes). Finally, if required, the projective alignment can be upgraded to a fully calibrated solution, as in the methods described above.

It is emphasized that the goal of this work is *not* to achieve the best possible photogrammetric reconstruction of the scene. Rather, the goal is to develop a practical way to associate color and texture information to each range point, as in Fig. 4.1. This output is intended to use in view-synthesis applications.

This chapter is organized as follows. Section 4.2.1 briefly reviews some standard material on projective reconstruction, while Sect. 4.2.2 describes the representation of range data in the present work. The chief contributions of the subsequent sections are as follows: Sect. 4.2.3 describes a *point-based* method that maps an ordinary *projective* reconstruction of the scene onto the corresponding range representation. This does not require the color cameras to be calibrated (although it may be necessary to correct for lens distortion). Any planar object can be used to find the alignment, provided that image features can be matched across all views (including that of the ToF camera). Section 4.2.4 describes a dual *plane-based* method, which performs the same projective alignment, but that does not require any point matches between the views. Any planar object can be used, provided that it has a simple polygonal boundary that can be segmented in the color and range data. This is a great advantage, owing to the very low resolution of the luminance data provided by the ToF camera (176×144 here). This makes it difficult to automatically extract and match point descriptors from these images, as described in Chap. 3. Furthermore, there are ToF devices that do not provide a luminance signal at all. Section 4.2.5 addresses the problem of multisystem alignment. Finally, Sect. 4.3 describes the accuracy than can be achieved with a three ToF+2RGB system, including a new error metric for ToF data in Sect. 4.3.2. Conclusions and future directions are discussed in Sect. 4.4.

4.2 Methods

This section describes the theory of projective alignment, using the following notation. Bold type will be used for vectors and matrices. In particular, points P, Q and planes U, V in the 3-D scene will be represented by column vectors of homogeneous coordinates, e.g.,

$$P = \begin{pmatrix} P_\Delta \\ P_4 \end{pmatrix} \quad \text{and} \quad U = \begin{pmatrix} U_\Delta \\ U_4 \end{pmatrix} \tag{4.1}$$

where $P_\Delta = (P_1, P_2, P_3)^\top$ and $U_\Delta = (U_1, U_2, U_3)^\top$. The homogeneous coordinates are defined up to a nonzero scaling; for example, $P \simeq (P_\Delta/P_4, 1)^\top$. In particular, if $P_4 = 1$, then P_Δ contains the ordinary space coordinates of the point P. Furthermore, if $|U_\Delta| = 1$, then U_4 is the signed perpendicular distance of the plane U from the origin, and U_Δ is the unit normal. The point P is on the plane U if $U^\top P = 0$. The cross-product $u \times v$ is often expressed as $(u)_\times v$, where $(u)_\times$ is a 3×3 antisymmetric matrix. The column vector of N zeros is written $\mathbf{0}_N$.

Projective cameras are represented by 3×4 matrices. For example, the range projection is

$$q \simeq CQ \quad \text{where} \quad C = \begin{pmatrix} A_{3\times 3} \mid b_{3\times 1} \end{pmatrix}. \tag{4.2}$$

The left and right color cameras C_ℓ and C_r are similarly defined, e.g., $p_\ell \simeq C_\ell P$. Table 4.1 summarizes the geometric objects that will be aligned.

Points and planes in the two systems are related by the unknown 4×4 space homography H, so that

4.2 Methods

Table 4.1 Summary of notations in the left, right, and range systems

	Observed Points	Reconstructed Points	Planes
Binocular C_ℓ, C_r	p_ℓ, p_r	P	U
Range C	(q, ρ)	Q	V

$$Q \simeq HP \quad \text{and} \quad V \simeq H^{-\top}U. \tag{4.3}$$

This model encompasses all rigid, similarity, and affine transformations in 3-D. It preserves *collinearity* and *flatness*, and is linear in homogeneous coordinates. Note that, in the reprojection process, H can be interpreted as a modification of the camera matrices, e.g., $p_\ell \simeq (C_\ell H^{-1})Q$, where $H^{-1}Q \simeq P$.

4.2.1 Projective Reconstruction

A projective reconstruction of the scene can be obtained from matched points $p_{\ell k}$ and p_{rk}, together with the fundamental matrix F, where $p_{rk}^\top F p_{\ell k} = 0$. The fundamental matrix can be estimated automatically, using the well-established RANSAC method. The camera matrices can then be determined, up to a four-parameter projective ambiguity [10]. In particular, from F and the epipole e_r, the cameras can be defined as

$$C_\ell \simeq (I \mid 0_3) \quad \text{and} \quad C_r \simeq \left((e_r)_\times F + e_r g^\top \mid \gamma e_r\right). \tag{4.4}$$

where $\gamma \neq 0$ and $g = (g_1, g_2, g_3)^\top$ can be used to bring the cameras into a plausible form. This makes it easier to visualize the projective reconstruction and, more importantly, can improve the numerical conditioning of subsequent procedures.

4.2.2 Range Fitting

The TOF camera C provides the distance ρ of each scene point from the camera center, as well as its image coordinates $q = (x, y, 1)$. The back projection of this point into the scene is

$$Q_\Delta = A^{-1}\left((\rho/\alpha) q - b\right) \quad \text{where} \quad \alpha = \left|A^{-1} q\right|. \tag{4.5}$$

Hence, the point $(Q_\Delta, 1)^\top$ is at distance ρ from the optical center $-A^{-1}b$, in the direction $A^{-1}q$. The scalar α serves to normalize the direction vector. This is the standard pinhole model, as used in [2].

The range data are noisy and incomplete, owing to illumination and scattering effects. This means that, given a sparse set of features in the intensity image (of the ToF device), it is not advisable to use the back-projected point (4.5) directly. A better approach is to segment the image of the plane in each ToF camera (using the range and/or intensity data). It is then possible to back project *all* of the enclosed points, and to robustly fit a plane V_j to the enclosed points Q_{ij}, so that $V_j^\top Q_{ij} \approx 0$ if point i lies on plane j. Now, the back projection Q_π of each sparse feature point q can be obtained by intersecting the corresponding ray with the plane V, so that the new range estimate ρ^π is

$$\rho^\pi = \frac{V_\Delta^\top A^{-1} b - V_4}{(1/\alpha) V_\Delta^\top A^{-1} q} \tag{4.6}$$

where $|V_4|$ is the distance of the plane to the camera center, and V_Δ is the unit normal of the range plane. The new point Q^π is obtained by substituting ρ^π into (4.5).

The choice of plane-fitting method is affected by two issues. First, there may be very severe outliers in the data, due to the photometric and geometric errors described in Chap. 1. Second, the noise-model should be based on the pinhole model, which means that perturbations occur radially along visual directions, which are not (in general) perpendicular to the observed plane [11, 24]. Several plane-fitting methods, both iterative [14] and noniterative [20], have been proposed for the pinhole model. The outlier problem, however, is often more significant. Hence, in practice, a RANSAC-based method is often the most effective.

4.2.3 Point-Based Alignment

It is straightforward to show that the transformation H in (4.3) could be estimated from five binocular points P_k, together with the corresponding range points Q_k. This would provide 5×3 equations, which determine the 4×4 entries of H, subject to an overall projective scaling. It is better, however, to use the 'Direct Linear Transformation' method [10], which fits H to *all* of the data. This method is based on the fact that if

$$P' = HP \tag{4.7}$$

is a perfect match for Q, then $\mu Q = \lambda P'$, and the scalars λ and μ can be eliminated between pairs of the four implied equations [5]. This results in $\binom{4}{2} = 6$ interdependent constraints per point. It is convenient to write these homogeneous equations as

$$\begin{pmatrix} Q_4 P'_\Delta - P'_4 Q_\Delta \\ Q_\Delta \times P'_\Delta \end{pmatrix} = \mathbf{0}_6. \tag{4.8}$$

Note that if P' and Q are normalized so that $P'_4 = 1$ and $Q_4 = 1$, then the magnitude of the top half of (4.8) is simply the distance between the points.

4.2 Methods

Following Förstner [7], the left-hand side of (4.8) can be expressed as $(Q)_\wedge P'$ where

$$(Q)_\wedge = \begin{pmatrix} Q_4 I_3 & -Q_\Delta \\ (Q_\Delta)_\times & 0_3 \end{pmatrix} \qquad (4.9)$$

is a 6×4 matrix, and $(Q_\Delta)_\times P_\Delta = Q_\Delta \times P_\Delta$, as usual. The Eq. (4.8) can now be written in terms of (4.7) and (4.9) as

$$(Q)_\wedge H P = 0_6. \qquad (4.10)$$

This system of equations is linear in the unknown entries of H, the columns of which can be stacked into the 16×1 vector h. The Kronecker product identity $\text{vec}(XYZ) = (Z^\top \otimes X) \text{vec}(Y)$ can now be applied, to give

$$\left(P^\top \otimes (Q)_\wedge\right) h = 0_6 \quad \text{where} \quad h = \text{vec}(H). \qquad (4.11)$$

If M points are observed on each of N planes, then there are $k = 1, \ldots, MN$ observed pairs of points, P_k from the projective reconstruction and Q_k from the range back projection. The MN corresponding 6×16 matrices $\left(P_k^\top \otimes (Q_k)_\wedge\right)$ are stacked together, to give the complete system

$$\begin{pmatrix} P_1^\top \otimes (Q_1)_\wedge \\ \vdots \\ P_{MN}^\top \otimes (Q_{MN})_\wedge \end{pmatrix} h = 0_{6MN} \qquad (4.12)$$

subject to the constraint $|h| = 1$, which excludes the trivial solution $h = 0_{16}$. It is straightforward to obtain an estimate of h from the SVD of the the $6MN \times 16$ matrix on the left of (4.12). This solution, which minimizes an *algebraic error* [10], is the singular vector corresponding to the smallest singular value of the matrix. In the minimal case, $M = 1, N = 5$, the matrix would be 30×16. Note that, the point coordinates should be transformed, to ensure that (4.12) is numerically well conditioned [10]. In this case, the transformation ensures that $\sum_k P_{k\Delta} = 0_3$ and $\frac{1}{MN} \sum_k |P_{k\Delta}| = \sqrt{3}$, where $P_{k4} = 1$. The analogous transformation is applied to the range points Q_k.

The DLT method, in practice, gives a good approximation H_{DLT} of the homography (4.3). This can be used as a starting point for the iterative minimization of a more appropriate error measure. In particular, consider the *reprojection error* in the left image,

$$E_\ell(C_\ell) = \sum_{k=1}^{MN} D(C_\ell Q_k, p_{\ell k})^2 \qquad (4.13)$$

where $D(p, q) = |p_\Delta/p_3 - q_\Delta/q_3|$. A 12-parameter optimization of (4.13), starting with $C_\ell \leftarrow C_\ell H_{\text{DLT}}^{-1}$, can be performed by the Levenberg-Marquardt algorithm [21].

The result will be the camera matrix $C_\ell^\star$ that best reprojects the range data into the left image ($C_r^\star$ is similarly obtained). The solution, provided that the calibration points adequately covered the scene volume, will remain valid for subsequent depth and range data.

Alternatively, it is possible to minimize the *joint* reprojection error, defined as the sum of left and right contributions,

$$E(H^{-1}) = E_\ell(C_\ell H^{-1}) + E_r(C_r H^{-1}) \qquad (4.14)$$

over the (inverse) homography H^{-1}. The 16 parameters are again minimized by the Levenberg-Marquardt algorithm, starting from the DLT solution H_{DLT}^{-1}.

The difference between the separate (4.13) and joint (4.14) minimizations is that the latter preserves the original epipolar geometry, whereas the former does not. Recall that C_ℓ C_r, H and F are all defined up to scale, and that F satisfies an additional rank-two constraint [10]. Hence, the underlying parameters can be counted as $(12-1) + (12-1) = 22$ in the separate minimizations, and as $(16-1) = 15$ in the joint minimization. The fixed epipolar geometry accounts for the $(9-2)$ missing parameters in the joint minimization. If F is known to be very accurate (or must be preserved) then the joint minimization (4.14) should be performed. This will also preserve the original binocular triangulation, provided that a projective-invariant method was used [9]. However, if minimal reprojection error is the objective, then the cameras should be treated separately. This will lead to a new fundamental matrix $F^\star = (e_r^\star)_\times C_r^\star(C_\ell^\star)^+$, where $(C_\ell^\star)^+$ is the generalized inverse. The right epipole is obtained from $e_r^\star = C_r^\star d_\ell^\star$, where $d_\ell^\star$ represents the nullspace $C_\ell^\star d_\ell^\star = 0_3$.

4.2.4 Plane-Based Alignment

The DLT algorithm of Sect. 4.2.3 can also be used to recover H from matched *planes*, rather than matched points. Equation (4.10) becomes

$$(V)_\wedge H^{-\top} U = 0_6 \qquad (4.15)$$

where U and V represent the estimated coordinates of the same plane in the parallax and range reconstructions, respectively. The estimation procedure is identical to that in Sect. 4.2.3, but with $\text{vec}(H^{-\top})$ as the vector of unknowns.

This method, in practice, produces very poor results. The chief reason that obliquely viewed planes are foreshortened, and therefore hard to detect/estimate, in the low-resolution ToF images. It follows that the calibration data set is biased towards fronto-parallel planes.[1] This bias allows the registration to slip sideways, perpendicular to the primary direction of the ToF camera. The situation is greatly

---

[1] The point-based algorithm is unaffected by this bias, because the scene is ultimately 'filled' with points, regardless of the contributing planes.

4.2 Methods

improved by assuming that the *boundaries* of the planes can be detected. For example, if the calibration object is rectangular, then the range projection of the plane V is bounded by four edges $\bar{v}_i$, where $i = 1, \ldots 4$. Note that, these are detected as *depth* edges, and so no luminance data are required. The edges, represented as lines $\bar{v}_i$, back project as the faces of a pyramid,

$$\overline{V}_i = C^\top \bar{v}_i = \begin{pmatrix} \overline{V}_{i\Delta} \\ 0 \end{pmatrix}, \quad i = 1, \ldots, L \qquad (4.16)$$

where $L = 4$ in the case of a quadrilateral projection. These planes are linearly dependent, because they pass through the center of projection; hence, the fourth coordinates are all zero if, as here, the ToF camera is at the origin. Next, if the corresponding edges $\bar{u}_{\ell i}$ and $\bar{u}_{r i}$ can be detected in the binocular system, using both color and parallax information, then the planes $\overline{U}_i$ can easily be constructed. Each calibration plane now contributes an additional $6L$ equations

$$(\overline{V}_i)_\wedge H^{-\top} \overline{U}_i = 0_6 \qquad (4.17)$$

to the DLT system (4.12). Although these equations are quite redundant (any two planes span all possibilities), they lead to a much better DLT estimate. This is because they represent exactly those planes that are most likely to be missed in the calibration data, owing to the difficulty of feature detection over surfaces that are extremely foreshortened in the image.

As in the point-based method, the plane coordinates should be suitably transformed, in order to make the numerical system (4.12) well conditioned. The transformed coordinates satisfy the location constraint $\sum_k U_{k\Delta} = 0_3$, as well as the scale constraint $\sum_k |U_{k\Delta}|^2 = 3 \sum_k U_{k4}^2$, where $U_{k\Delta} = (U_{k1}, U_{k2}, U_{k3})^\top$, as usual. A final renormalization $|U_k| = 1$ is also performed. This procedure, which is also applied to the V_k, is analogous to the treatment of line coordinates in DLT methods [26].

The remaining problem is that the original reprojection error (4.13) cannot be used to optimize the solution, because no luminance features q have been detected in the range images (and so no 3-D points Q have been distinguished). This can be solved by reprojecting the physical edges of the calibration planes, after reconstructing them as follows. Each edge plane $\overline{V}_i$ intersects the range plane V in a space-line, represented by the 4×4 Plücker matrix

$$W_i = V \overline{V}_i^\top - \overline{V}_i V^\top. \qquad (4.18)$$

The line W_i reprojects to a 3×3 antisymmetric matrix [10]; for example

$$W_{\ell i} \simeq C_\ell W_i C_\ell^\top \qquad (4.19)$$

in the left image, and similarly in the right. Note that $W_{\ell i} p_\ell = 0$ if the point p_ℓ is on the reprojected line [10]. The line-reprojection error can therefore be written as

$$E_\ell^\times(C_\ell) = \sum_{i=1}^{L}\sum_{j=1}^{N} D_\times(C_\ell W_i C_\ell^\top, \bar{u}_{\ell i j})^2. \qquad (4.20)$$

The function $D_\times(M, n)$ compares image lines, by computing the sine of the angle between the two coordinate vectors,

$$D_\times(M, n) = \frac{\sqrt{2}\,|Mn|}{|M|\,|n|} = \frac{|m \times n|}{|m|\,|n|}, \qquad (4.21)$$

where $M = (m)_\times$, and $|M|$ is the Frobenius norm. It is emphasized that the coordinates *must* be normalized by a suitable transformations G_ℓ and G_r, as in the case of the DLT. For example, the line n should be fitted to points of the form Gp, and then M should be transformed as $G^{-\top}M$, before computing (4.21). The reprojection error (4.20) is numerically unreliable without this normalization.

The line reprojection (4.21) can either be minimized separately for each camera, or jointly as

$$E^\times(H^{-1}) = E_\ell^\times(C_\ell H^{-1}) + E_r^\times(C_r H^{-1}) \qquad (4.22)$$

by analogy with (4.14). Finally, it should be noted that although (4.21) is defined in the *image*, it is an *algebraic* error. However, because the errors in question are small, this measure behaves predictably (see Fig. 4.2).

4.2.5 Multisystem Alignment

The point-based and plane-based procedures, described in Sects. 4.2.3 and 4.2.4 respectively, can be used to calibrate a single ToF+2RGB system. Related methods can be used for the joint calibration of several such systems, as will now be explained, using the *point-based* representation. In this section, the notation P_i will be used for the binocular coordinates (with respect to the left camera) of a point in the i-th system, and likewise Q_i for the ToF coordinates of a point in the same system. Hence, the i-th ToF, left and right RGB cameras have the form

$$C_i \simeq (A_i \mid 0_3), \quad C_{\ell i} \simeq (A_{\ell i} \mid 0_3) \quad \text{and} \quad C_{ri} \simeq (A_{ri} \mid b_{ri}) \qquad (4.23)$$

where A_i and $A_{\ell i}$ contain only *intrinsic* parameters, whereas A_{ri} also encodes the relative orientation of C_{ri} with respect to $C_{\ell i}$. Each system has a transformation H_i^{-1} that maps ToF points Q_i into the corresponding RGB coordinate system of $C_{\ell i}$. Furthermore, let the 4 × 4 matrix G_{ij} be the transformation from system j, mapping *back* to system i. This matrix, in the calibrated case, would be a rigid 3-D transformation. However, by analogy with the ToF-to-RGB matrices, each G_{ij} is generalized here to a projective transformation, thereby allowing for spatial distortions in the data. The left and right cameras that project a scene point P_j in coordinate system j

4.2 Methods

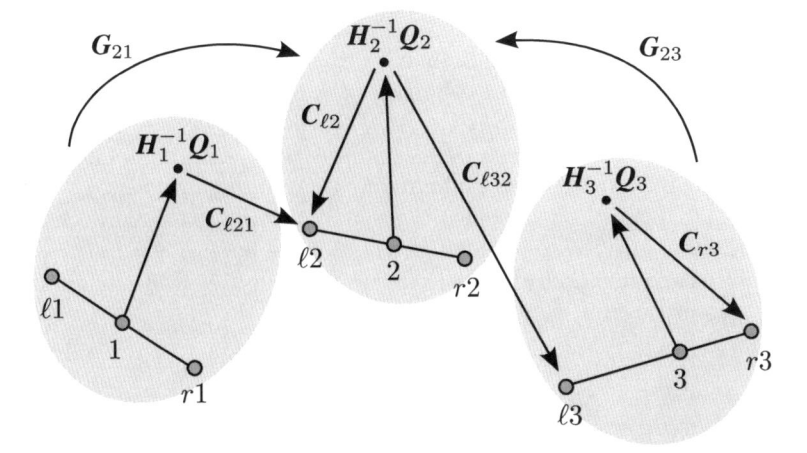

Fig. 4.3 Example of a three ToF+2RGB setup, with ToF cameras labeled 1,2,3. Each *ellipse* represents a separate system, with system 2 chosen as the reference. The *arrows* (with camera-labels) show some possible ToF-to-RGB projections. For example, a point $P_2 \simeq H_2^{-1} Q_2$ in the center projects directly to RGB view $\ell 2$ via $C_{\ell 2}$, whereas the same point projects to $\ell 3$ via $C_{\ell 32} = C_{\ell 3} G_{32}$

to image points $p_{\ell i}$ and p_{ri} in system i are

$$C_{\ell ij} = C_{\ell i}\, G_{ij} \quad \text{and} \quad C_{rij} = C_{ri}\, G_{ij}. \qquad (4.24)$$

Note that if a single global coordinate system is chosen to coincide with the k-th RGB system, then a point P_k projects via $C_{\ell ik}$ and C_{rik}. These two cameras are respectively equal to $C_{\ell i}$ and C_{ri} in (4.23) only when $i = k$, such that $G_{ij} = I$ in (4.24). A typical three-system configuration is shown in Fig. 4.3.

The transformation G_{ij} can only be estimated directly if there is a region of common visibility between systems i and j. If this is not the case (as when the systems face each other, such that the front of the calibration board is not simultaneously visible), then G_{ij} can be computed indirectly. For example, $G_{02} = G_{01}\, G_{12}$ where $P_2 = G_{12}^{-1} G_{01}^{-1} P_0$. Note that, the stereo-reconstructed points P are used to estimate these transformations, as they are more reliable than the ToF points Q.

4.3 Evaluation

The following sections will describe the accuracy of a nine-camera setup, calibrated by the methods described above. Section 4.3.1 will evaluate *calibration* error, whereas Sect. 4.3.2 will evaluate *total* error. The former is essentially a fixed function of the estimated camera matrices, for a given scene. The latter also includes the range noise from the ToF cameras, which varies from moment to moment. The importance of this distinction will be discussed.

The setup consists of three rail-mounted ToF+2RGB systems, $i = 1\ldots 3$, as in Fig. 4.3. The stereo baselines are 17 cm on average, and the ToF cameras are separated by 107 cm on average. The RGB images are 1624×1224, whereas the Mesa Imaging SR4000 ToF images are 176×144, with a depth range of 500 cm. The three stereo systems are first calibrated by standard methods, returning a full Euclidean decomposition of $\boldsymbol{C}_{\ell i}$ and $\boldsymbol{C}_{ri}$, as well as the associated lens parameters. It was established in [8] that projective alignment is generally superior to similarity alignment, and so the transformations $\boldsymbol{G}_{ij}$ and $\boldsymbol{H}_j^{-1}$ will be 4×4 homographies. These transformations were estimated by the DLT method, and refined by LM-minimization of the joint geometric error, as in (4.14).

4.3.1 Calibration Error

The calibration error is measured by first taking ToF points $\boldsymbol{Q}_j^\pi$ corresponding to *vertices* on the reconstructed calibration plane π_j in system j, as described in Sect. 4.2.2. These can then be projected into a pair of RGB images in system i, so that the error $E_{ij}^{\text{cal}} = \frac{1}{2}\bigl(E_{\ell ij}^{\text{cal}} + E_{rij}^{\text{cal}}\bigr)$ can be computed, where

$$E_{\ell ij}^{\text{cal}} = \frac{1}{|\pi|} \sum_{\boldsymbol{Q}_j^\pi} D\bigl(\boldsymbol{C}_{\ell ij}\, \boldsymbol{H}_j^{-1}\, \boldsymbol{Q}_j^\pi,\ \boldsymbol{p}_{\ell i}\bigr) \qquad (4.25)$$

and E_{rij}^{cal} is similarly defined. The function $D(\cdot, \cdot)$ computes the image distance between inhomogenized points, as in (4.13), and the denominator corresponds to the number of vertices on the board, with $|\pi| = 35$ in the present experiments. The measure (4.25) can of course be averaged over all images in which the board is visible. The calibration procedure has an accuracy of around 1 pixel, as shown in Fig. 4.4.

4.3.2 Total Error

The calibration error, as reported in the preceding section, is the natural way to evaluate the estimated cameras and homographies. It is not, however, truly representative of the 'live' performance of the complete setup. This is because the calibration error uses each estimated plane π_j to replace all vertices $\boldsymbol{Q}_j$ with the *fitted* versions $\boldsymbol{Q}_j^\pi$. In general, however, no surface model is available, and so the raw points $\boldsymbol{Q}_j$ must be used as input for meshing and rendering processes.

The total error, which combines the calibration and range errors, can be measured as follows. The i-th RGB views of plane π_j must be related to the ToF image points $\boldsymbol{q}_j$ by the 2-D *transfer* homographies $\boldsymbol{T}_{\ell ij}$ and $\boldsymbol{T}_{rij}$, where

4.3 Evaluation

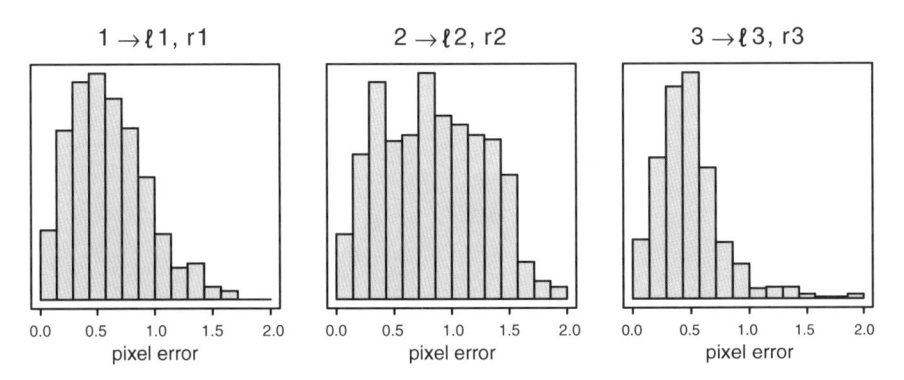

Fig. 4.4 Calibration error (4.25), measured by projecting the fitted ToF points Q^π to the left and right RGB images (1624 × 1224) in three separate systems. Each histogram combines *left*-camera and *right*-camera measurements from 15 views of the calibration board. Subpixel accuracy is obtained

$$p_{\ell i} \simeq T_{\ell ij} q_j \quad \text{and} \quad p_{ri} \simeq T_{rij} q_j. \tag{4.26}$$

These 3 × 3 matrices can be estimated accurately, because the range data itself is not required. Furthermore, let Π_j be the hull (i.e., bounding polygon) of plane π_j as it appears in the ToF image. Any pixel q_j in the hull (including the original calibration vertices) can now be *reprojected* to the i-th RGB views via the 3-D point Q_j, or *transferred* directly by $T_{\ell ij}$ and T_{rij} in (4.26). The total error is the average difference between the reprojections and the transfers, $E^{\text{tot}}_{ij} = \frac{1}{2}(E^{\text{tot}}_{\ell ij} + E^{\text{tot}}_{rij})$, where

$$E^{\text{tot}}_{\ell ij} = \frac{1}{|\Pi_j|} \sum_{q_j \in \Pi_j} D\left(C_{\ell ij} H_j^{-1} Q_j, \ T_{\ell ij} q_j\right) \tag{4.27}$$

and E^{tot}_{rij} is similarly defined. The view-dependent denominator $|\Pi_j| \gg |\pi|$ is the number of pixels in the hull Π_j. Hence, E^{tot}_{ij} is the total error, including range noise, of ToF plane π_j as it appears in the i-th RGB cameras.

If the RGB cameras are not too far from the ToF camera, then the range errors tend to be canceled in the reprojection. This is evident in Fig. 4.5, although it is clear that the tail of each distribution is increased by the range error. However, if the RGB cameras belong to another system, with a substantially different location, then the range errors can be very large in the reprojection. This is clear from Fig. 4.6, which shows that a substantial proportion of the ToF points reproject to the other systems with a total error in excess of 10 pixels.

It is possible to understand these results more fully by examining the distribution of the total error across individual boards. Figure 4.7 shows the distribution for a board reprojected to the same system (i.e., part of the data from Fig. 4.5). There is a relatively smooth gradient of error across the board, which is attributable to errors in the fitting of plane π_j, and in the estimation of the camera parameters. The pixels

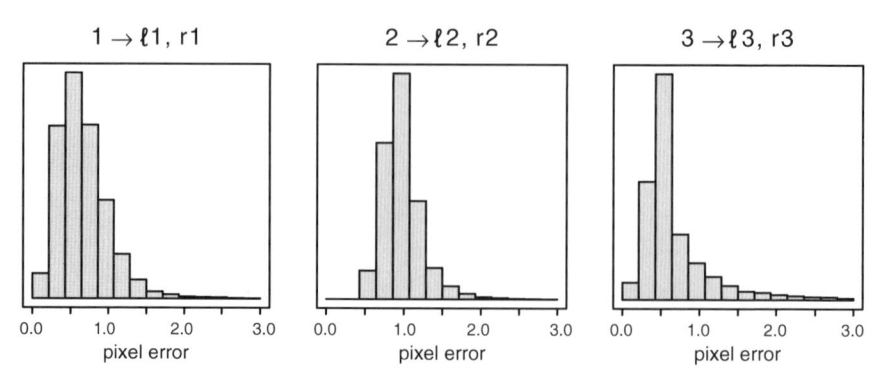

Fig. 4.5 Total error (4.27), measured by projecting the raw ToF points Q to the left and right RGB images (1624 × 1224) in three separate systems. These distributions have longer and heavier tails than those of the corresponding calibration errors, shown in Fig. 4.4

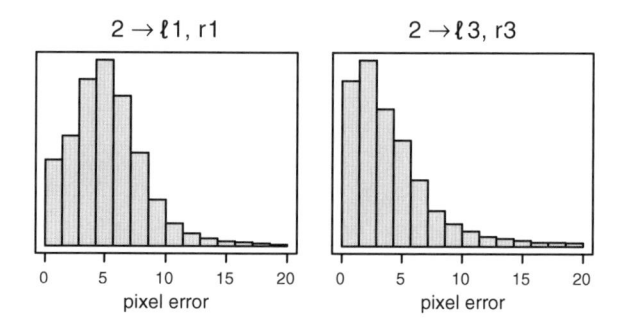

Fig. 4.6 Total error when reprojecting raw ToF points from system 2 to RGB cameras in systems 1 and 3 (*left and right*, respectively). The range errors are emphasized by the difference in viewpoints between the two systems. Average error is now around 5 pixels in the 1624 × 1224 images, and the noisiest ToF points reproject with tens of pixels of error

can be divided into sets from the black and white squares, using the known board geometry and detected vertices. It can be seen in Fig. 4.7 (right) that the total error for each set is comparable. However, when reprojecting to a different system, Fig. 4.8 shows that the total error is correlated with the black and white squares on the board. This is due to significant absorption of the infrared signal by the black squares.

4.4 Conclusions

It has been shown that there is a projective relationship between the data provided by a ToF camera, and an uncalibrated binocular reconstruction. Two practical methods for computing the projective transformation have been introduced; one that requires luminance point correspondences between the ToF and color cameras, and one that

4.4 Conclusions

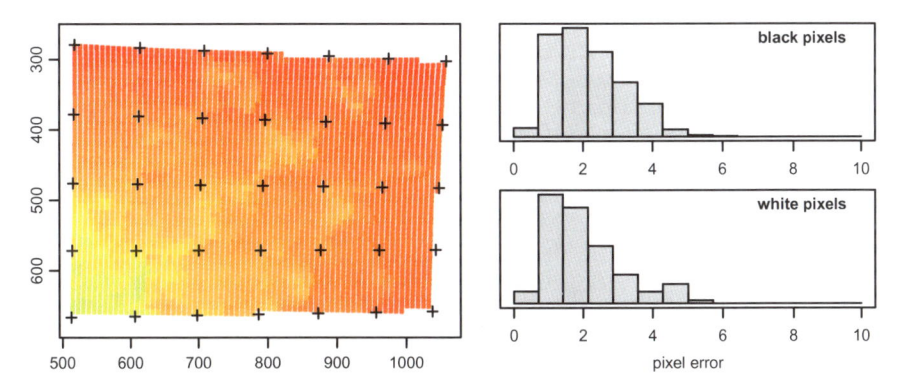

Fig. 4.7 *Left* 3-D TOF pixels ($|\Pi| = 3216$), on a calibration board, reprojected to an RGB image in the same TOF+2RGB system. Each pixel is color coded by the total error (4.27). *Black crosses* are the detected vertices in the RGB image. *Right* histograms of total error, split into pixels on *black* or *white squares*

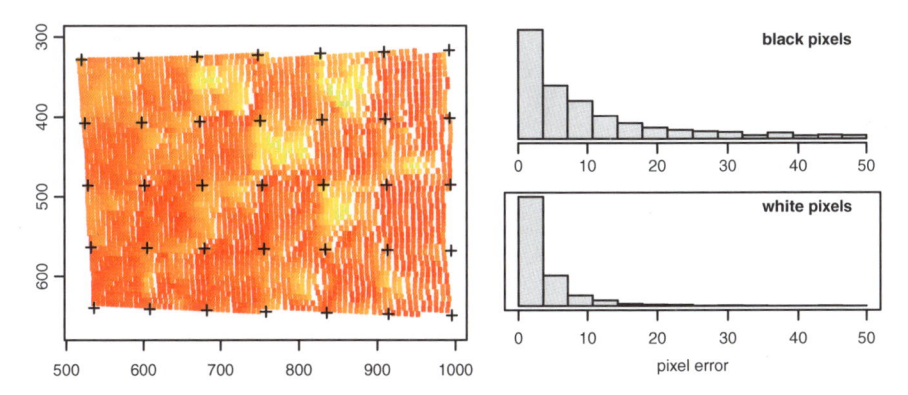

Fig. 4.8 *Left* 3-D TOF pixels, as in Fig. 4.7, reprojected to an RGB image in a different TOF+2RGB system. *Right* histograms of total error, split into pixels on *black or white squares*. The depth of the *black squares* is much less reliable, which leads to inaccurate reprojection into the target system

does not. Either of these methods can be used to associate binocular color and texture with each 3-D point in the range reconstruction. It has been shown that the point-based method can easily be extended to multiple-TOF systems, with calibrated or uncalibrated RGB cameras.

The problem of TOF noise, especially when reprojecting 3-D points to a very different viewpoint, has been emphasized. This source of error can be reduced by application of the denoising methods described in Chap. 1. Alternatively, having aligned the TOF and RGB systems, it is possible to refine the 3-D representation by image matching, as explained in Chap. 5.

References

1. Bartczak, B., Koch, R.: Dense depth maps from low resolution time-of-flight depth and high resolution color views. In: Proceedings of International Symposium on Visual Computing (ISVC), pp. 228–239 (2009)
2. Beder, C., Bartczak, B., Koch, R.: A comparison of PMD-cameras and stereo-vision for the task of surface reconstruction using patchlets. In: Proceedings of Computer Vision and Parallel Recognition (CVPR), pp. 1–8 (2007)
3. Beder, C., Schiller, I., Koch, R.: Photoconsistent relative pose estimation between a PMD 2D3D-camera and multiple intensity cameras. In: Proceedings of Symposium of the German Association for Pattern Recognition (DAGM), pp. 264–273 (2008)
4. Bleiweiss, A., Werman, M.: Fusing time-of-flight depth and color for real-time segmentation and tracking. In: Proceedings of the Dynamic 3D Imaging: DAGM 2009 Workshop, pp. 58–69 (2009)
5. Csurka, G., Demirdjian, D., Horaud, R.: Finding the collineation between two projective reconstructions. Comput. Vis. Image Underst. **75**(3), 260–268 (1999)
6. Dubois, J.M., Hügli, H.: Fusion of time-of-flight camera point clouds. In: Proceedings of European Conference on Computer Vision (ECCV) Workshop on Multi-Camera and Multimodal Sensor Fusion Algorithms and Applications, Marseille (2008)
7. Förstner, W.: Uncertainty and projective geometry. In: Bayro-Corrochano, E. (ed.) Handbook of Geometric Computing, pp. 493–534. Springer, New York (2005)
8. Hansard, M., Horaud, R., Amat, M., Lee, S.: Projective alignment of range and parallax data. In: Proceedings of Computer Vision and Parallel Recognition (CVPR), pp. 3089–3096 (2011)
9. Hartley, R., Sturm, P.: Triangulation. Comput. Vis. Image Underst. **68**(2), 146–157 (1997)
10. Hartley, R., Zisserman, A.: Multiple View Geometry in Computer Vision. Cambridge University Press, Cambridge (2000)
11. Hebert, M., Krotkov, E.: 3D measurements from imaging laser radars: how good are they? Image Vis. Comput. **10**(3), 170–178 (1992)
12. Horn, B., Hilden, H., Negahdaripour, S.: Closed-form solution of absolute orientation using orthonormal matrices. J. Opt. Soc. Am. A **5**(7), 1127–1135 (1988)
13. Huhle, B., Fleck, S., Schilling, A.: Integrating 3D time-of-flight camera data and high resolution images for 3DTV applications. In: Proceedings of 3DTV Conference, pp. 1–4 (2007)
14. Kanazawa, Y., Kanatani, K.: Reliability of plane fitting by range sensing. In: International Conference on Robotics and Automation (ICRA), pp. 2037–2042 (1995)
15. Kim, Y., Chan, D., Theobalt, C., Thrun, S.: Design and calibration of a multi-view TOF sensor fusion system. In: Proceedings of Computer Vision and Parallel Recognition (CVPR) Workshop on Time-of-Flight Camera based Computer Vision (2008)
16. Koch, R., Schiller, I., Bartczak, B., Kellner, F., Köser, K.: MixIn3D: 3D mixed reality with ToF-camera. In: Proceedings of DAGM Workshop on Dynamic 3D Imaging, pp. 126–141 (2009)
17. Kolb, A., Barth, E., Koch, R., Larsen, R.: Time-of-flight cameras in computer graphics. Comput. Graphics Forum **29**(1), 141–159 (2010)
18. Lindner, M., Schiller, I., Kolb, A., Koch, R.: Time-of-flight sensor calibration for accurate range sensing. Comput. Vis. Image Underst. **114**(12), 1318–1328 (2010)
19. Mesa Imaging AG. http://www.mesa-imaging.ch
20. Pathak, K., Vaskevicius, N., Birk, A.: Revisiting uncertainty analysis for optimum planes extracted from 3D range sensor point-clouds. In: Proceedings of IEEE International Conference on Robotics and Automation (ICRA), pp. 1631–1636 (2009)
21. Press, W.H., Teukolsky, S.A., Vetterling, W.T., Flannery, B.P.: Numerical Recipes in C. Cambridge University Press, 2nd edition (1992)
22. Schiller, I., Beder, C., Koch, R.: Calibration of a PMD camera using a planar calibration object together with a multi-camera setup. In: International Archives of Photogrammetry, Remote Sensing and Spatial Information Sciences XXI, pp. 297–302 (2008)

References

23. Verri, A., Torre, V.: Absolute depth estimate in stereopsis. J. Opt. Soc. Am. A **3**(3), 297–299 (1986)
24. Wang, C., Tanahasi, H., Hirayu, H., Niwa, Y., Yamamoto, K.: Comparison of local plane fitting methods for range data. In: Proceedings of Computer Vision and Parallel Recognition (CVPR), pp. 663–669 (2001)
25. Wu, J., Zhou, Y., Yu, H., Zhang, Z.: Improved 3D depth image estimation algorithm for visual camera. In: Proceedings of International Congress on Image and Signal Processing (2009)
26. Zeng, H., Deng, X., Hu, Z.: A new normalized method on line-based homography estimation. Pattern Recogn. Lett. **29**, 1236–1244 (2008)
27. Zhang Q., Pless, R.: Extrinsic calibration of a camera and laser range finder (improves camera calibration). In: Proceedings of Intenational Conference on Intelligent Robots and Systems, pp. 2301–2306 (2004)
28. Zhu, J., Wang, L., Yang, R.G., Davis, J.: Fusion of time-of-flight depth and stereo for high accuracy depth maps. In: Proceedings of Computer Vision and Parallel Recognition (CVPR), pp. 1–8 (2008)

Chapter 5
A Mixed Time-of-Flight and Stereoscopic Camera System

Abstract Several methods that combine range and color data have been investigated and successfully used in various applications. Most of these systems suffer from the problems of noise in the range data and resolution mismatch between the range sensor and the color cameras. High-resolution depth maps can be obtained using stereo matching, but this often fails to construct accurate depth maps of weakly/repetitively textured scenes. Range sensors provide coarse depth information regardless of presence/absence of texture. We propose a novel ToF-stereo fusion method based on an efficient seed-growing algorithm which uses the ToF data projected onto the stereo image pair as an initial set of correspondences. These initial "seeds" are then propagated to nearby pixels using a matching score that combines an image similarity criterion with rough depth priors computed from the low-resolution range data. The overall result is a dense and accurate depth map at the resolution of the color cameras at hand. We show that the proposed algorithm outperforms 2D image-based stereo algorithms and that the results are of higher resolution than off-the-shelf RGB-D sensors, e.g., Kinect.

Keywords Mixed-camera systems · Stereo seed-growing · Time-of-Flight sensor fusion · Depth and color combination

5.1 Introduction

Advanced computer vision applications require both depth and color information. Hence, a system composed of ToF and color cameras should be able to provide accurate *color and depth* information for each pixel and at high resolution. Such a *mixed* system can be very useful for a large variety of vision problems, e.g., for building dense 3D maps of indoor environments.

The 3D structure of a scene can be reconstructed from two or more 2D views via a *parallax* between corresponding image points. However, it is difficult to obtain accurate pixel-to-pixel matches for scenes of objects without textured surfaces, with repetitive patterns, or in the presence of occlusions. The main drawback is that stereo matching algorithms frequently fail to reconstruct indoor scenes composed of untextured surfaces, e.g., walls, repetitive patterns and surface discontinuities, which are typical in man-made environments.

Alternatively, *active-light* range sensors, such as time-of-flight (ToF) or structured-light cameras (see Chap. 1), can be used to directly measure the 3D structure of a scene at video frame rates. However, the spatial resolution of currently available range sensors is lower than high-definition (HD) color cameras, the luminance sensitivity is poorer and the depth range is limited. The range-sensor data are often noisy and incomplete over extremely scattering parts of the scene, e.g., non-Lambertian surfaces. Therefore, it is not judicious to rely solely on range-sensor estimates for obtaining 3D maps of complete scenes. Nevertheless, range cameras provide good initial estimates independently of whether the scene is textured or not, which is not the case with stereo matching algorithms. These considerations show that it is useful to combine the active-range and the passive-parallax approaches, in a *mixed* system. Such a system can overcome the limitations of both the active- and passive-range (stereo) approaches, when considered separately, and provides accurate and fast 3D reconstruction of a scene at high resolution, e.g., 1200×1600 pixels, as in Fig. 5.1.

5.1.1 Related Work

The combination of a depth sensor with a color camera has been exploited in several applications such as object recognition [2, 15, 24], person awareness, gesture recognition [11], simultaneous localization and mapping (SLAM) [3, 17], robotized plant-growth measurement [1], etc. These methods mainly deal with the problem of noise in depth measurement, as examined in Chap. 1, as well as with the low resolution of range data as compared to the color data. Also, most of these methods are limited to RGB-D, i.e., a *single* color image combined with a range sensor. Interestingly enough, the recently commercialized Kinect [13] camera falls in the RGB-D family of sensors. We believe that extending the RGB-D sensor model to RGB-D-RGB sensors is extremely promising and advantageous because, unlike the former type of sensor, the latter type can combine active depth measurement with stereoscopic matching and hence better deal with the problems mentioned above.

Stereo matching has been one of the most studied paradigms in computer vision. There are several papers, e.g., [22, 23] that overview existing techniques and that highlight recent progress in stereo matching and stereo reconstruction. While a detailed description of existing techniques is beyond the scope of this section, we note that algorithms based on greedy local search techniques are typically fast but frequently fail to reconstruct the poorly textured regions or ambiguous surfaces. Alternatively, global methods formulate the matching task as an optimization problem which leads the minimization of a Markov random field (MRF) energy function of the image similarity likelihood and a prior on the surface smoothness.

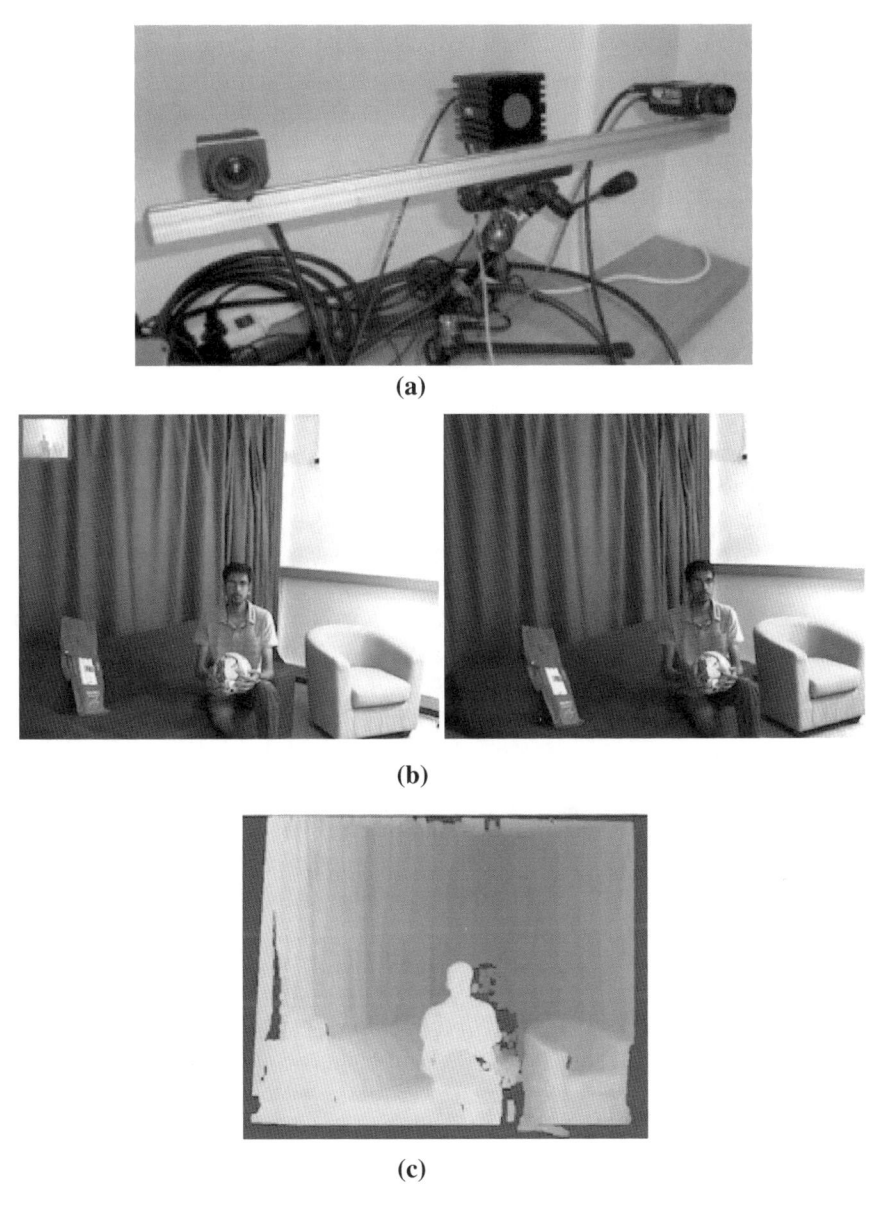

5.1 Introduction

Fig. 5.1 **a** Two high-resolution color cameras (2.0 MP at 30 FPS) are combined with a single low-resolution ToF camera (0.03 MP at 30 FPS). **b** The 144 × 177 ToF image (*upper left corner*) and two 1224 × 1624 *color* images are shown at the true scale. **c** The depth map obtained with our method. The technology used by both these camera types allows simultaneous range and photometric data acquisition with an extremely accurate temporal synchronization, which may not be the case with other types of range cameras such as the current version of Kinect

These algorithms solve some of the aforementioned problems of local methods but are very complex and computationally expensive since optimizing an MRF-based energy function is an NP-hard problem in the general case.

A practical tradeoff between the local and the global methods in stereo is the seed-growing class of algorithms [4–6]. The correspondences are grown from a small set of initial correspondence seeds. Interestingly, they are not particularly sensitive to bad input seeds. They are significantly faster than the global approaches, but they have difficulties in presence of nontextured surfaces; Moreover, in these cases they yield depth maps which are relatively sparse. Denser maps can be obtained by relaxing the matching threshold but this leads to erroneous growth, so there is a natural tradeoff between the accuracy and density of the solution. Some form of regularization is necessary in order to take full advantage of these methods.

Recently, external prior-based generative probabilistic models for stereo matching were proposed [14, 20] for reducing the matching ambiguities. The priors used were based on surface triangulation obtained from an initially matched distinctive interest points in the two color images. Again, in the absence of textured regions, such support points are only sparsely available, and are not reliable enough or are not available at all in some image regions, hence the priors are erroneous. Consequently, such prior-based methods produce artifacts where the priors win over the data, and the solution is biased toward such incorrect priors. This clearly shows the need for more accurate prior models. Wang et al. [25] integrate a regularization term based on the depth values of initially matched *ground control points* in a global energy minimization framework. The ground control points are gathered using an accurate laser scanner. The use of a laser scanner is tedious because it is difficult to operate and because it cannot provide depth measurements fast enough such that it can be used in a practical computer vision application.

ToF cameras are based on an active sensor principle[1] that allows 3D data acquisition at video frame rates, e.g., 30 FPS as well as accurate synchronization with any number of color cameras.[2] A modulated infrared light is emitted from the camera's internal lighting source, is reflected by objects in the scene and eventually travels back to the sensor, where the time of flight between sensor and object is measured independently at each of the sensor's pixel by calculating the precise phase delay between the emitted and the detected waves. A complete depth map of the scene can thus be obtained using this sensor at the cost of very low spatial resolution and coarse depth accuracy (see Chap. 1 for details).

The fusion of ToF data with stereo data has been recently studied. For example, [8] obtained a higher quality depth map, by a probabilistic ad hoc fusion of ToF and stereo data. Work in [26] merges the depth probability distribution function obtained from ToF and stereo. However, both these methods are meant for improvement over the initial data gathered with the ToF camera and the final depth-map result is still limited to the resolution of the ToF sensor. The method proposed in this chapter

---

[1] All experiments described in this chapter use the Mesa SR4000 camera [18].
[2] http://www.4dviews.com

5.1 Introduction

increases the resolution from 0.03 MP to the full resolution of the color cameras being used, e.g., 2 MP.

The problem of depth-map upsampling has been also addressed in the recent past. In [7] a noise-aware filter for adaptive multilateral upsampling of ToF depth maps is presented. The work described in [15, 21] extends the model of [9], and [15] demonstrates that the object detection accuracy can be significantly improved by combining a state-of-the-art 2D object detector with 3D depth cues. The approach deals with the problem of resolution mismatch between range and color data using an MRF-based superresolution technique in order to infer the depth at every pixel. The proposed method is slow: It takes around 10 s to produce a 320×240 depth image. All of these methods are limited to depth-map upsampling using only a single color image and do not exploit the added advantage offered by stereo matching, which can highly enhance the depth map both qualitatively and quantitatively. Recently, [12] proposed a method which combines ToF estimates with stereo in a semiglobal matching framework. However, at pixels where ToF disparity estimates are available, the image similarity term is ignored. This make the method quite susceptible to errors in regions where ToF estimates are not precise, especially in textured regions where stereo itself is reliable.

5.1.2 Chapter Contributions

In this chapter, we propose a novel method for incorporating range data within a robust seed-growing algorithm for stereoscopic matching [4]. A calibrated system composed of an active-range sensor and a stereoscopic color camera pair, as described in Chap. 4 and [16], allows the range data to be aligned and then projected onto each one of the two images, thus providing an initial sparse set of point-to-point correspondences (seeds) between the two images. This initial seed set is used in conjunction with the seed-growing algorithm proposed in [4]. The projected ToF points are used as the vertices of a mesh-based surface representation which, in turn, is used as a prior to regularize the image-based matching procedure. The novel probabilistic *fusion* model proposed here (between the mesh-based surface initialized from the sparse ToF data and the seed-growing stereo matching algorithm itself) combines the merits of the two 3D sensing methods (active and passive) and overcomes some of the limitations outlined above. Notice that the proposed fusion model can be incorporated within virtually any stereo algorithm that is based on energy minimization and which requires some form initialization. It is, however, particularly efficient and accurate when used in combination with match-propagation methods.

The remainder of this chapter is structured as follows: Sect. 5.2 describes the proposed range-stereo fusion algorithm. The growing algorithm is summarized in Sect. 5.2.1. The processing of the ToF correspondence seeds is explained in Sect. 5.2.2, and the sensor fusion based similarity statistic is described in Sect. 5.2.3. Experimental results on a real data set and evaluation of the method, are presented in Sect. 5.3. Finally, Sect. 5.4 draws some conclusions.

5.2 The Proposed ToF-Stereo Algorithm

As outlined above, the TOF camera provides a low-resolution depth map of a scene. This map can be projected onto the left and right images associated with the stereoscopic pair, using the projection matrices estimated by the calibration method described in Chap. 4. Projecting a single 3D point (x, y, z) gathered by the TOF camera onto the *rectified* images provides us with a pair of corresponding points (u, v) and (u', v') with $v' = v$ in the respective images. Each element (u, u', v) denotes a point in the disparity space.[3] Hence, projecting all the points obtained with the TOF camera gives us a sparse set of 2D point correspondences. This set is termed as the set of initial support points or TOF *seeds*.

These initial support points are used in a variant of the seed-growing stereo algorithm [4, 6] which further grows them into a denser and higher resolution disparity map. The seed-growing stereo algorithms propagate the correspondences by searching in the small neighborhoods of the seed correspondences. Notice that this growing process limits the disparity space to be visited to only a small fraction, which makes the algorithm extremely efficient from a computational point-of-view. The limited neighborhood also gives a kind of implicit regularization, nevertheless the solution can be arbitrarily complex, since multiple seeds are provided.

The integration of range data within the seed-growing algorithm requires two major modifications: (1) The algorithm is using TOF seeds instead of the seeds obtained by matching distinctive image features, such as interest points, between the two images, and (2) the growing procedure is regularized using a similarity statistic which takes into account the photometric consistency as well as the depth likelihood based on disparity estimate by interpolating the rough triangulated TOF surface. This can be viewed as a prior cast over the disparity space.

5.2.1 The Growing Procedure

The growing algorithm is sketched in pseudocode as Algorithm 1. The input is a pair of rectified images (I_L, I_R), a set of *refined* TOF seeds $\mathscr{S}$ (see below), and a parameter τ which directly controls a tradeoff between matching accuracy and matching density. The output is a disparity map D which relates pixel correspondences between the input images.

First, the algorithm computes the prior disparity map D_p by interpolating TOF seeds. Map D_p is of the same size as the input images and the output disparity map, Step 1. Then, a similarity statistic simil $(s|I_L, I_R, D_p)$ of the correspondence, which measures both the photometric consistency of the potential correspondence as well as its consistency with the prior, is computed for all seeds $s = (u, u', v) \in \mathscr{S}$, Step 2. Recall that the seed s stands for a pixel-to-pixel correspondence $(u, v) \leftrightarrow (u', v)$

---

[3] The disparity space is a space of all potential correspondences [22].

5.2 The Proposed ToF-Stereo Algorithm

Algorithm 1 Growing algorithm for ToF-stereo fusion

Require: Rectified images (I_L, I_R),
initial correspondence seeds $\mathscr{S}$,
image similarity threshold τ.

1: Compute the prior disparity map D_p by interpolating seeds $\mathscr{S}$.
2: Compute $\mathrm{simil}(s|I_L, I_R, D_p)$ for every seed $s \in \mathscr{S}$.
3: Initialize an empty disparity map D of size I_L (and D_p).
4: **repeat**
5:    Draw seed $s \in \mathscr{S}$ of the best $\mathrm{simil}(s|I_L, I_R, D_p)$ value.
6:    **for** each of the four best neighbors $i \in \{1, 2, 3, 4\}$
$$q_i^* = (u, u', v) = \underset{q \in \mathscr{N}_i(s)}{\mathrm{argmax}}\ \mathrm{simil}(q|I_L, I_R, D_p)$$
   do
7:      $c := \mathrm{simil}(q_i^*|I_L, I_R, D_p)$
8:      **if** $c \geq \tau$ **and** pixels not matched yet **then**
9:         Update the seed queue $\mathscr{S} := \mathscr{S} \cup \{q_i^*\}$.
10:        Update the output map $D(u, v) = u - u'$.
11:      **end if**
12:    **end for**
13: **until** $\mathscr{S}$ is empty
14: **return** disparity map D.

between the left and the right images. For each seed, the algorithm searches other correspondences in the surroundings of the seeds by maximizing the similarity statistic. This is done in a 4-neighborhood $\{\mathscr{N}_1, \mathscr{N}_2, \mathscr{N}_3, \mathscr{N}_4\}$ of the pixel correspondence, such that in each respective direction (left, right, up, down) the algorithm searches the disparity in a range of ± 1 pixel from the disparity of the seed, Step 6. If the similarity statistic of a candidate exceeds the threshold value τ, then a new correspondence is found, Step 8. This new correspondence becomes itself a new seed, and the output disparity map D is updated accordingly. The process repeats until there are no more seeds to be grown.

The algorithm is robust to a fair percentage of wrong initial seeds. Indeed, since the seeds compete to be matched based on a best-first strategy, the wrong seeds typically have low score $\mathrm{simil}(s)$ associated with them and therefore when they are evaluated in Step 5, it is likely that the involved pixels been already matched. For more details on the growing algorithm, we refer the reader to [4, 6].

5.2.2 ToF Seeds and Their Refinement

The original version of the seed-growing stereo algorithm [6] uses an initial set of seeds $\mathscr{S}$ obtained by detecting interest points in both images and matching them. Here, we propose to use ToF seeds. As already outlined, these seeds are obtained by projecting the low-resolution depth map associated with the ToF camera onto the high-resolution images. Likewise the case of interest points, this yields a sparse set of seeds, e.g., approximately 25,000 seeds in the case of the ToF camera used in

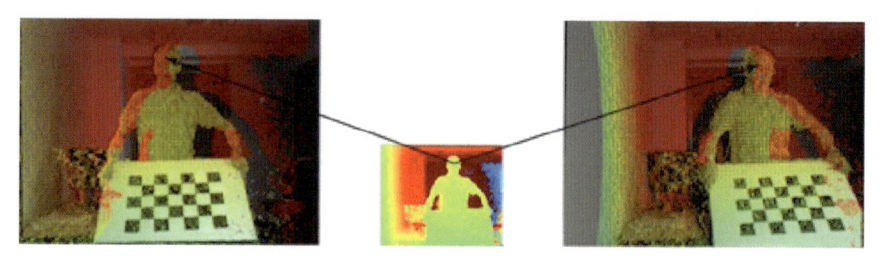

Fig. 5.2 This figure shows an example of the projection of the ToF points onto the left and right images. The projected points are color coded such that the color represents the disparity: cold colors correspond to large disparity values. Notice that there are many wrong correspondences on the computer monitor due to the screen reflectance and to artifacts along the occlusion boundaries

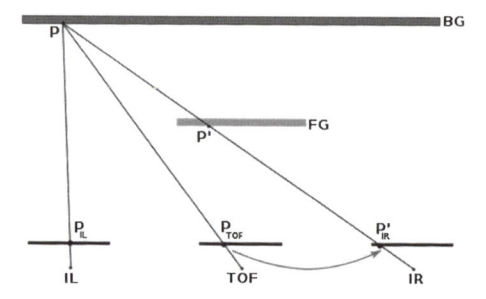

Fig. 5.3 The effect of occlusions. A ToF point P that belongs to a background (BG) objects is only observed in the left image (IL), while it is occluded by a foreground object (FG), and hence not seen in the right image (IR). When the ToF point P is projected onto the *left* and *right* images, an incorrect correspondence ($P_{IL} \leftrightarrow P'_{IR}$) is established

our experiments. Nevertheless, one of the main advantages of the ToF seeds over the interest points is that they are regularly distributed across the images regardless of the presence/absence of texture. This is not the case with interest points whose distribution strongly depends on texture as well as lighting conditions, etc. Regularly distributed seeds will provide a better coverage of the observed scene, i.e., even in the absence of textured areas.

However, ToF seeds are not always reliable. Some of the depth values associated with the ToF sensor are inaccurate. Moreover, whenever a ToF point is projected onto the left and onto the right images, it does not always yield a valid stereo match. There may be several sources of error which make the ToF seeds less reliable than one would have expected, as in Figs. 5.2 and 5.3. In detail:

1. *Imprecision due to the calibration process.* The transformations allowing to project the 3D ToF points onto the 2D images are obtained via a complex sensor calibration process, i.e., Chap. 4. This introduces localization errors in the image planes of up to 2 pixels.

5.2 The Proposed ToF-Stereo Algorithm

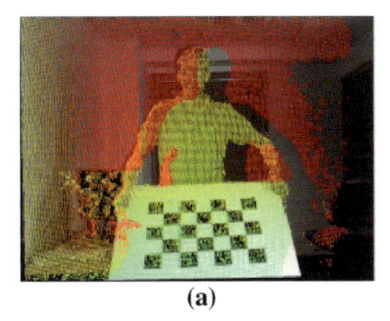

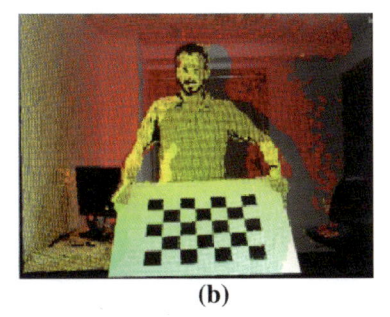

Fig. 5.4 An example of the effect of correcting the set of seeds on the basis that they should be regularly distributed. **a** Original set of seeds. **b** Refined set of seeds

2. *Outliers due to the physical/geometric properties of the scene.* Range sensors are based on active light and on the assumption that the light beams travel from the sensor and back to it. There are a number of situations where the beam is lost, such as specular surfaces, absorbing surfaces (such as fabric), scattering surfaces (such as hair), slanted surfaces, bright surfaces (computer monitors), faraway surfaces (limited range), or when the beam travels in an unpredictable way, such a multiple reflections.
3. *The ToF camera and the 2D cameras observe the scene from slightly different points of view.* Therefore, it may occur that a 3D point that is present in the ToF data is only seen into the left or right image, as in Fig. 5.3, or is not seen at all.

Therefore, a fair percentage of the ToF seeds are *outliers*. Although the seed-growing stereo matching algorithm is robust to the presence of outliers in the initial set of seeds, as already explained in Sect. 5.2.1, we implemented a straightforward refinement step in order to detect and eliminate incorrect seed data, prior to applying Algorithm 1. First, the seeds that lie in low-intensity (very dark) regions are discarded since the ToF data are not reliable in these cases. Second, in order to handle the background-to-foreground occlusion effect just outlined, we detect seeds which are not uniformly distributed across image regions. Indeed, projected 3D points lying on smooth fronto-parallel surfaces form a regular image pattern of seeds, while projected 3D points that belong to a background surface and which project onto a foreground image region do not form a regular pattern, e.g., occlusion boundaries in Fig. 5.4a.

Nonregular seed patterns are detected by counting the seed occupancy within small 5×5 pixel windows around every seed point in both images. If there is more than one seed point in a window, the seeds are classified as belonging to the background and hence they are discarded. A refined set of seeds is shown in Fig. 5.4b. The refinement procedure typically filters 10–15 % of all seed points.

5.2.3 Similarity Statistic Based on Sensor Fusion

The original seed-growing matching algorithm [6] uses Moravec's normalized cross-correlation [19] (MNCC),

$$\text{simil}(s) = \text{MNCC}(w_L, w_R) = \frac{2\text{cov}(w_L, w_R)}{\text{var}(w_L) + \text{var}(w_R) + \varepsilon} \quad (5.1)$$

as the similarity statistic to measure the photometric consistency of a correspondence $s : (u, v) \leftrightarrow (u', v)$. We denote by w_L and w_R the feature vectors which collect image intensities in small windows of size $n \times n$ pixels centered at (u, v) and $(u'v)$ in the left and right image, respectively. The parameter ε prevents instability of the statistic in cases of low-intensity variance. This is set as the machine floating point epsilon. The statistic has low response in textureless regions and therefore the growing algorithm does not propagate the correspondences across these regions. Since the ToF sensor can provide seeds without the presence of any texture, we propose a novel similarity statistic, $\text{simil}(s|I_L, I_R, D_p)$. This similarity measure uses a different score for photometric consistency as well as an initial high-resolution disparity map D_p, both incorporated into the Bayesian model explained in detail below.

The initial disparity map D_p is computed as follows. A 3D meshed surface is built from a 2D triangulation applied to the ToF image. The disparity map D_p is obtained via interpolation from this surface such that it has the same (high) resolution as of the left and right images. Figure 5.5a, b show the meshed surface projected onto the left high-resolution image and built from the ToF data, before and after the seed refinement step, which makes the D_p map more precise.

Let us now consider the task of finding an optimal high-resolution disparity map. For each correspondence $(u, v) \leftrightarrow (u', v)$ and associated disparity $d = u - u'$ we seek an optimal disparity d^* such that:

$$d^* = \underset{d}{\text{argmax}} \ P(d|I_L, I_R, D_p). \quad (5.2)$$

By applying the Bayes' rule, neglecting constant terms, assuming that the distribution $P(d)$ is uniform in a local neighborhood where it is sought (Step. 6), and considering conditional independence $P(I_l, I_r, D|d) = P(I_L, I_R|d)P(D_p|d)$, we obtain:

$$d^* = \underset{d}{\text{argmax}} \ P(I_L, I_R|d)P(D_p|d), \quad (5.3)$$

where the first term is the color-image likelihood and the second term is the range-sensor likelihood. We define the color-image and range-sensor likelihoods as:

$$P(I_L, I_R|d) \propto \text{EXPSSD}(w_L, w_R)$$
$$= \exp\left(-\frac{\sum_{i=1}^{n \times n}(w_L(i) - w_R(i))^2}{\sigma_s^2 \sum_{i=1}^{n \times n}(w_L(i)^2 + w_R(i)^2)}\right), \quad (5.4)$$

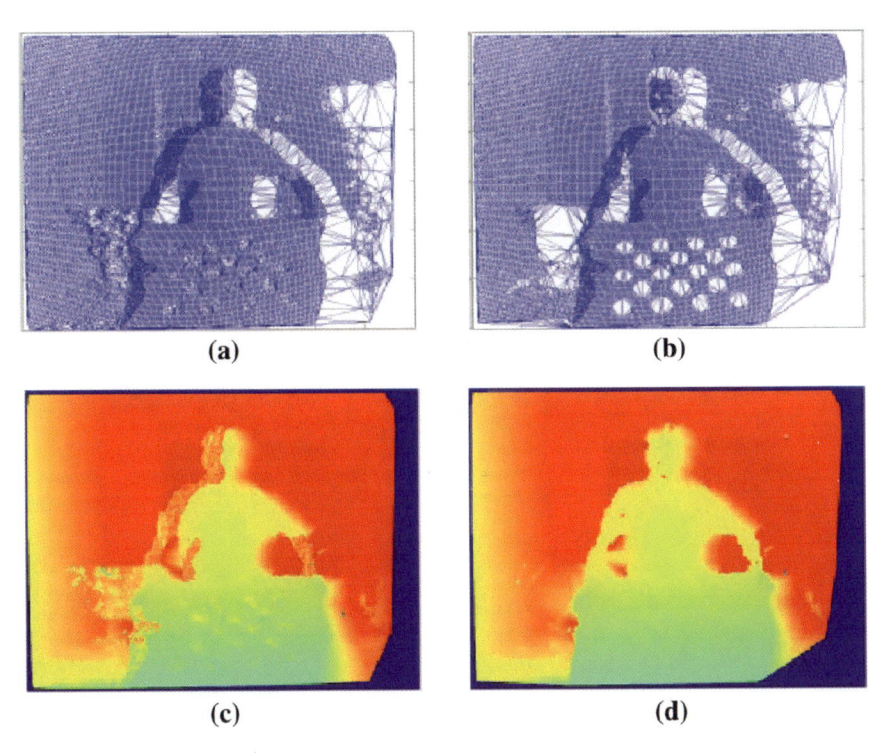

Fig. 5.5 Triangulation and prior disparity map D_p. These are shown using both raw seeds **a**, **c** and refined seeds **b**, **d**. A positive impact of the refinement procedure is clearly visible

and as:

$$P(D_p|d) \propto \exp\left(-\frac{(d-d_p)^2}{2\sigma_p^2}\right) \tag{5.5}$$

respectively, where σ_s are σ_p two normalization parameters. Therefore, the new similarity statistic becomes:

$$\begin{aligned}\text{simil}(s|I_L, I_R, D_p) &= \text{EPC}(w_L, w_R, D_p)\\ &= \exp\left(-\frac{\sum_{i=1}^{n\times n}(w_L(i)-w_R(i))^2}{\sigma_s^2 \sum_{i=1}^{n\times n}(w_L(i)^2+w_R(i)^2)} - \frac{(d-d_p)^2}{2\sigma_p^2}\right).\end{aligned} \tag{5.6}$$

Notice that the proposed image likelihood has a high response for correspondences associated with textureless regions. However, in such regions, all possible matches have similar image likelihoods. The proposed range-sensor likelihood regularizes the solution and forces it toward the one closest to the prior disparity map D_p. A tradeoff between these two terms can be obtained by tuning the parameters σ_s and σ_p. We refer to this similarity statistic as the *exponential prior correlation* (EPC) score.

5.3 Experiments

Our experimental setup comprises one Mesa Imaging SR4000 ToF camera [18] and a pair of high-resolution Point Grey[4] color cameras, as shown in Fig. 5.1. The two color cameras are mounted on a rail with a baseline of about 49 cm and the ToF camera is approximately midway between them. All three optical axes are approximately parallel. The resolution of the ToF image is of 144 × 176 pixels and the color cameras have a resolution of 1224 × 1624 pixels. Recall that Fig. 5.1b highlights the resolution differences between the ToF and color images. This camera system was calibrated using the alignment method of Chap. 4.

In all our experiments, we set the parameters of the method as follows: Windows of 5×5 pixels were used for matching ($n = 5$), the matching threshold in Algorithm 1 is set to $\tau = 0.5$, the balance between the photometric and range-sensor likelihoods is governed by two parameters in (5.6), which were set to $\sigma_s^2 = 0.1$ and to $\sigma_p^2 = 0.001$.

We show both qualitatively and quantitatively (using data sets with ground truth) the benefits of the range sensor and an impact of particular variants of the proposed fusion model integrated in the growing algorithm. Namely, we compare results of (i) the original stereo algorithm [6] with MNCC correlation and Harris seeds (MNCC-Harris), (ii) the same algorithm with ToF seeds (MNCC-TOF), (iii) the algorithm which uses EXPSSD similarity statistic instead with both Harris (EXPSSD-Harris) and ToF seeds (EXPSSD-TOF), and (iv) the full sensor fusion model of the regularized growth (EPC). Finally, small gaps of unassigned disparity in the disparity maps were filled by a primitive procedure which assigns median disparity in the 5×5 window around the gap (EPC—gaps filled). These small gaps usually occur in slanted surfaces, since Algorithm 1 in Step. 8 enforces one-to-one pixel matching. Nevertheless this way, they can be filled easily, if needed.

5.3.1 Real-Data Experiments

We captured two real-world data sets using the camera setup described above, SET-1 in Fig. 5.6 and SET-2 in Fig. 5.7. Notice that in both of these examples the scene surfaces are weakly textured. Results shown as disparity maps are color coded, such that warmer colors are further away from the cameras and unmatched pixels are dark blue.

In Fig. 5.6d, we can see that the original algorithm [6] has difficulties in weakly textured regions which results in large unmatched regions due to the MNCC statistic (5.1), and it produces several mismatches over repetitive structures on the background curtain, due to erroneous (mismatched) Harris seeds. In Fig. 5.6e, we can see that after replacing the sparse and somehow erratic Harris seeds with uniformly distributed (mostly correct) ToF seeds, the results have significantly been improved. There are no more mismatches on the background, but unmatched regions are still large. In Fig. 5.6f, the EXPSSD statistic (5.4) was used instead of MNCC which

---

[4] http://www.ptgrey.com/

5.3 Experiments

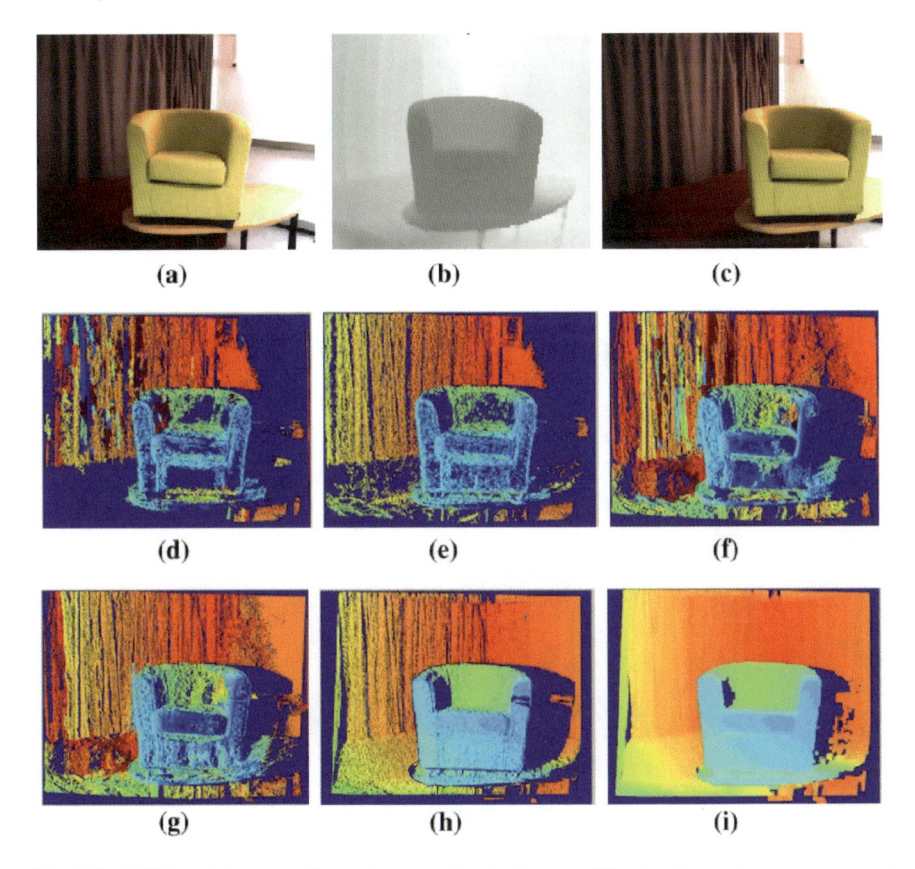

Fig. 5.6 SET-1: **a** left image, **b** TOF image and **c** right image. The TOF image has been zoomed at the resolution of the color images for visualization purposes. Results obtained **d** using the seed-growing stereo algorithm [6] combining Harris seeds and MNCC statistic, **e** using TOF seeds and MNCC statistic, **f** using Harris seeds and EXPSSD statistic, **g** using TOF seeds with EXPSSD statistics. Results obtained with the proposed stereo-TOF fusion model using the EPC (exponential prior correlation) similarity statistic **h**, and EPC after filling small gaps **i**

causes similar mismatches as in Fig. 5.6d, but unlike MNCC there are matches in textureless regions, nevertheless mostly erratic. The reason is that unlike MNCC statistic the EXPSSD statistic has high response in low-textured regions. However, since all disparity candidates have equal (high) response inside such regions, the unregularized growth is random, and produces mismatches. The situation does not improve much using the TOF seeds, as shown in Fig. 5.6g. Significantly better results are finally shown in Fig. 5.6h which uses the proposed EPC fusion model EPC from Eq. (5.6). The EPC statistic, unlike EXPSSD, has the additional regularizing range-sensor likelihood term which guides the growth in ambiguous regions and attracts the solution toward the initial depth estimates of the TOF camera. Results are further refined by filling small gaps, as shown in Fig. 5.6i. Similar observations can be made in Fig. 5.7. The proposed model clearly outperforms the other discussed approaches.

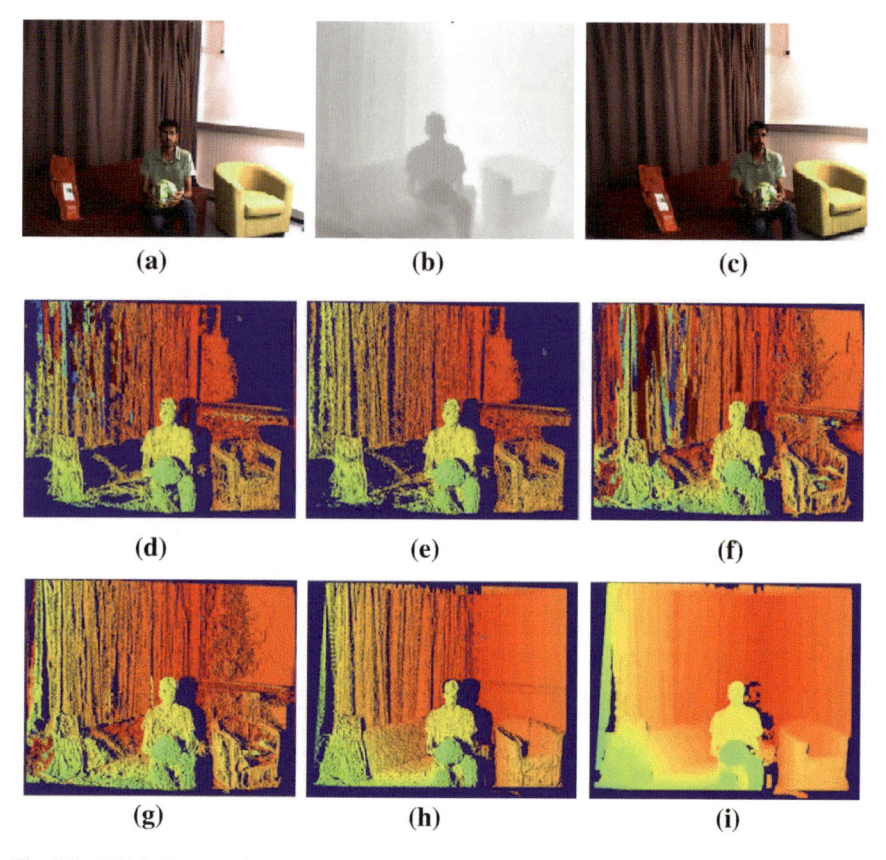

Fig. 5.7 SET-2. Please refer to the caption of Fig. 5.6 for explanations. **a** Left image. **b** ToF image (*zoomed*). **c** Right image. **d** MNCC-Harris. **e** MNCC-ToF. **f** EXPSSD-Harris. **g** EXPSSD-ToF. **h** EPC (proposed). **i** EPC (gaps filled)

5.3.2 Comparison Between ToF Map and Estimated Disparity Map

For the proper analysis of a stereo matching algorithm, it is important to inspect the reconstructed 3D surfaces. Indeed, the visualization of the disparity/depth maps can sometimes be misleading. Surface reconstruction reveals fine details in the quality of the results. This is in order to qualitatively show the gain of the high-resolution depth map produced by the proposed algorithm with respect to the low-resolution depth map of the ToF sensor.

In order to provide a fair comparison, we show the reconstructed surfaces associated with the *dense* disparity maps D_p obtained after 2D triangulation of the ToF data points, Fig. 5.8a, as well as the reconstructed surfaces associated with the disparity map obtained with the proposed method, Fig. 5.8b. Clearly, much more of the sur-

5.3 Experiments

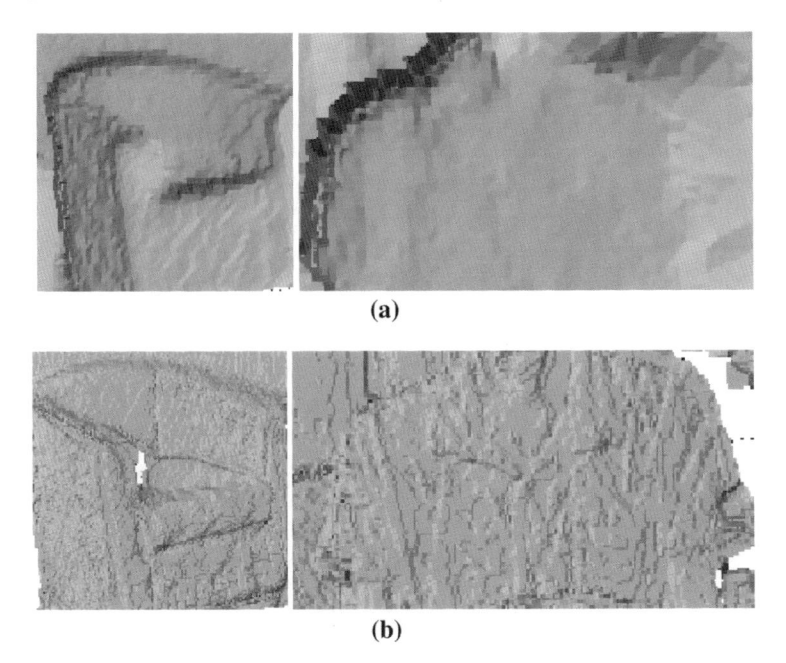

Fig. 5.8 The reconstructed surfaces are shown as relighted 3D meshes for **a** the prior disparity map D_p (2D triangulation on projected and refined ToF seeds), and **b** for the disparity map obtained using the proposed algorithm. Notice the fine surface details which were recovered by the proposed method

face details are recovered by the proposed method. Notice precise object boundaries and fine details, like the cushion on the sofa chair and a collar of the T-shirt, which appear in Fig. 5.8b. This qualitatively corroborates the precision of the proposed method compared to the ToF data.

5.3.3 Ground-Truth Evaluation

To quantitatively demonstrate the validity of the proposed algorithm, we carried out an experiment on data sets with associated ground-truth results. Similarly to [8] we used the Middlebury data set [22] and simulated the ToF camera by sampling the ground-truth disparity map.

The following results are based on the Middlebury-2006 data set.[5] On purpose, we selected three challenging scenes with weakly textured surfaces: Lampshade-1, Monopoly, Plastic. The input images are of size 1330×1110 pixels. We took every 10th pixel in a regular grid to simulate the ToF camera. This gives us about 14k of ToF points, which is roughly the same ratio to color images as for the real sensors. We are aware that simulation ToF sensor this way is naive, since we do not simulate

---

[5] http://vision.middlebury.edu/stereo/data/scenes2006/

any noise or artifacts, but we believe that for validating the proposed method this is satisfactory.

Results are shown in Fig. 5.9 and Table 5.1. We show the left input image, results of the same algorithms as in the previous section with the real sensor, and the ground-truth disparity map. For each disparity, we compute the percentage of correctly matched pixels in nonoccluded regions. This error statistic is computed as the number of pixels for which the estimated disparity differs from the ground-truth disparity by less than one pixel, divided by number of all pixels in nonoccluded regions. Notice that, unmatched pixels are considered as errors of the same kind as mismatches. This is in order to allow a strict but fair comparison between algorithms which deliver solutions of different densities. The quantitative evaluation confirms the previous observations regarding the real-world setup. The proposed algorithm, which uses the full sensor fusion model, significantly outperforms all other tested variants.

For the sake of completeness, we also report error statistics for the prior disparity map D_p which is computed by interpolating ToF seeds, see step 1 of Algorithm 1. These are 92.9, 92.1, 96.0 % for Lampshade-1, Monopoly, Plastic scene, respectively. These results are already quite good, which means the interpolation we use to construct the prior disparity map is appropriate. These scenes are mostly piecewise planar, which the interpolation captures well. On the other hand, recall that in the real case, not all the seeds are correct due to various artifacts of the range data. Nevertheless in all three scenes, the proposed algorithm (EPC with gaps filled) was able to further improve the precision up to 96.4, 95.3, 98.2 % for the respective scenes. This is again consistent with the experiments with the real ToF sensor, where higher surface details were recovered, see Fig. 5.8.

5.3.4 Computational Costs

The original growing algorithm [6] has low computational complexity due to intrinsic search space reduction. Assuming the input stereo images are of size $n \times n$ pixels, the algorithm has the complexity of $\mathcal{O}(n^2)$, while any exhaustive algorithm has the complexity at least $\mathcal{O}(n^3)$ as noted in [5]. The factor n^3 is the size of the search space in which the correspondences are sought, i.e., the disparity space. The growing algorithm does not compute similarity statistics of all possible correspondences, but efficiently traces out components of high similarity score around the seeds. This low complexity is beneficial especially for high-resolution imagery, which allows precise surface reconstruction.

The proposed algorithm with all presented modifications does not represent any significant extra cost. Triangulation of ToF seeds and the prior disparity map computation is not very costly, and nor is computation of the new EPC statistic (instead of MNCC).

For our experiments, we use an "academic", i.e., a combined Matlab/C implementation which takes approximately 5 s on two million pixel color images. An efficient implementation of the seed-growing algorithm [6] which runs in real time

5.3 Experiments

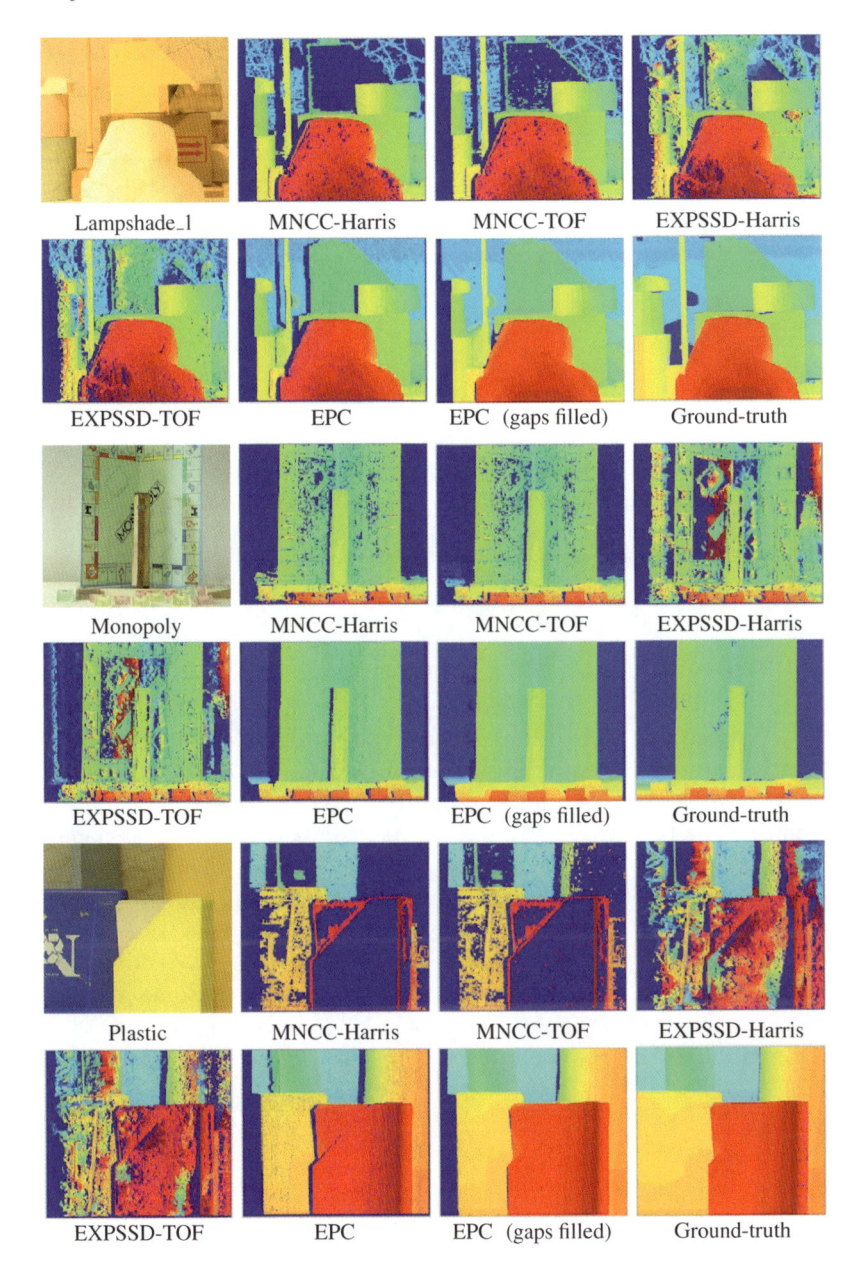

Lampshade_1	MNCC-Harris	MNCC-TOF	EXPSSD-Harris
EXPSSD-TOF	EPC	EPC (gaps filled)	Ground-truth
Monopoly	MNCC-Harris	MNCC-TOF	EXPSSD-Harris
EXPSSD-TOF	EPC	EPC (gaps filled)	Ground-truth
Plastic	MNCC-Harris	MNCC-TOF	EXPSSD-Harris
EXPSSD-TOF	EPC	EPC (gaps filled)	Ground-truth

Fig. 5.9 Middlebury data set. *Left-right* and *top-bottom*: the left images, results obtained with the same algorithms as in Figs. 5.6 and 5.7, and the ground-truth disparity maps. This evaluation shows that the combination of the proposed seed-growing stereo algorithm with a prior disparity map, obtained from a sparse and regularly distributed set of 3D points, yields excellent dense matching results

Table 5.1 The error statistics (percentage of correctly matched pixels) associated with the tested algorithms and for three test image pairs from the Middlebury data set

Left image	MNCC-Harris(%)	MNCC-TOF(%)	EXPSSD-Harris(%)	EXPSSD-TOF(%)	EPC(%)	EPC (gaps filled)(%)
Lampshade-1	61.5	64.3	44.9	49.5	88.8	96.4
Monopoly	51.2	53.4	29.4	32.1	85.2	95.3
Plastic	25.2	28.2	13.5	20.6	88.7	98.2

on a standard CPU was recently proposed [10]. This indicates that a real-time implementation of the proposed algorithm is feasible. Indeed, the modification of the growing algorithm and integration with the ToF data does not bring any significant extra computational costs. The algorithmic complexity remains the same, since we have only slightly modified the similarity score used inside the growing procedure. It is true that prior to the growing process, the ToF data must be triangulated. Nevertheless, this can be done extremely efficiently using computer graphics techniques and associated software libraries.

5.4 Conclusions

We have proposed a novel correspondence growing algorithm, performing fusion of a range sensor and a pair of passive color cameras, to obtain an accurate and dense 3D reconstruction of a given scene. The proposed algorithm is robust, and performs well on both textured and textureless surfaces, as well as on ambiguous repetitive patterns. The algorithm exploits the strengths of the ToF sensor and those of stereo matching between color cameras, in order to compensate for their individual weaknesses. The algorithm has shown promising results on difficult real-world data, as well as on challenging standard data sets which quantitatively corroborates its favorable properties. Together with the strong potential for real-time performance that has been discussed, the algorithm would be practically very useful in many computer vision and robotic applications.

References

1. Alenyà, G., Dellen, B., Torras, C.: 3D Modelling of leaves from color and tof data for robotized plant measuring. In: Proceedings of IEEE International Conference on Robotics and Automation (ICRA), pp. 3408–3414 (2011)
2. Attamimi, M., Mizutani, A., Nakamura, T., Nagai, T., Funakoshi, K., Nakano, M.: Real-time 3D visual sensor for robust object recognition. In: Proceedings of IEEE/RSJ International Conference on Intelligent Robots and Systems (IROS), pp. 4560–4565 (2010)
3. Castañeda, V., Mateus, D., Navab, N.: SLAM combining ToF and high-resolution cameras. In: IEEE Workshkop on Motion and Video Computing (2011)

References

4. Cech, J., Matas, J., Perdoch, M.: Efficient sequential correspondence selection by cosegmentation. IEEE Trans. Pattern Anal. Mach. Intell. **32**(9), 1568–1581 (2010)
5. Cech, J., Sanchez-Riera, J., Horaud, R.: Scene flow estimation by growing correspondence seeds. In: Proceedings of the IEEE Conference on Computer Vision and Pattern Recognition (CVPR), pp. 3129–3136 (2011)
6. Čech, J., Šára, R.: Efficient sampling of disparity space for fast and accurate matching. In: Proceedings of BenCOS Workshop CVPR (2007)
7. Chan, D., Buisman, H., Theobalt, C., Thrun, S.: A noise-aware filter for real-time depth upsampling. In: ECCV Workshop on Multi-camera and Multi-modal Sensor Fusion Algorithms and Applications (2008)
8. Dal Mutto, C., Zanuttigh, P., Cortelazzo, G.M.: A probabilistic approach to ToF and stereo data fusion. In: Proceedings of 3D Data Processing, Visualization and Transmission, Paris (2010)
9. Diebel, J., Thrun, S.: An application of Markov random fields to range sensing. In: Proceedings on Neural Information Processing Systems (NIPS) (2005)
10. Dobias, M., Sára, R.: Real-time global prediction for temporally stable stereo. In: Proceedings of IEEE International Conference on Computer Vision Workshops, pp. 704–707 (2011)
11. Droeschel, D., Stückler, J., Holz, D., Behnke, S.: Towards joint attention for a domestic service robot—person awareness and gesture recognition using time-of-flight cameras. In: Proceedings of International Conference on Robotic and Animation (ICRA), Shanghai, pp. 1205–1210 (2011)
12. Fischer, J., Arbeiter, G., Verl, A.: Combination of time-of-flight depth and stereo using semiglobal optimization. In: Proceedings of IEEE International Conference on Robotics and Automation (ICRA), pp. 3548–3553 (2011)
13. Freedman, B., Shpunt, A., Machline, M., Arieli, Y.: Depth Mapping Using Projected Patterns. US Patent No. 8150412 (2012)
14. Geiger, A., Roser, M., Urtasun, R.: Efficient large-scale stereo matching. In: Proceedings of the Asian Conference on Computer Vision (ACCV), pp. 25–38 (2010)
15. Gould, S., Baumstarck, P., Quigley, M., Ng, A.Y., Koller, D.: Integrating visual and range data for robotic object detection. In: Proceedings of European Conference on Computer Vision Workshops (2008)
16. Hansard, M., Horaud, R., Amat, M., Lee, S.: Projective alignment of range and parallax data. In: Proceedings of IEEE Conference on Computer Vision and Pattern Recognition (CVPR), pp. 3089–3096 (2011)
17. Jebari, I., Bazeille, S., Battesti, E., Tekaya, H., Klein, M., Tapus, A., Filliat, D., Meyer, C., Sio-Hoi, I., Benosman, R., Cizeron, E., Mamanna, J.-C., Pothier, B.: Multi-sensor semantic mapping and exploration of indoor environments. In: Technologies for Practical Robot Applications (TePRA), pp. 151–156 (2011)
18. Mesa Imaging AG. http://www.mesa-imaging.ch
19. Moravec, H.P.: Toward automatic visual obstacle avoidance. In: 5th International Conference Artificial Intelligence (ICAI), pp. 584–594 (1977)
20. Newcombe, R.A., Davison, A.J.: Live dense reconstruction with a single moving camera. In: Proceedings of IEEE Conference on Computer Vision and Pattern Recognition (CVPR), pp. 1498–1505 (2010)
21. Park, J., Kim, H., Tai, Y.-W., Brown, M.-S., Kweon, I.S.: High quality depth map upsampling for 3D-TOF cameras. In: Proceedings of IEEE International Conference on Computer Vision (ICCV) (2011)
22. Scharstein, D., Szeliski, R.: A taxonomy and evaluation of dense two-frame stereo correspondence algorithms. Int. J. Comput. Vision **47**, 7–42 (2002)
23. Seitz, S., Curless, B., Diebel, J., Scharstein, D., Szeliski, R.: A comparison and evaluation of multi-view stereo reconstruction algorithms. In: Proceedings of Computer Vision and Pattern Recognition (CVPR), pp. 519–528 (2006)
24. Stückler J., Behnke, S.: Combining depth and color cues for scale- and viewpoint-invariant object segmentation and recognition using random forests. In: Proceedings of IEEE/ RSJ International Conference on Robots and Systems (IROS), pp. 4566–4571 (2010)

25. Wang, L., Yang, R.: Global stereo matching leveraged by sparse ground control points. In: Proceedings of IEEE Conference on Computer Vision and Pattern Recognition (CVPR), pp. 3033–3040 (2011)
26. Zhu, J., Wang, L., Yang, R.G., Davis, J.: Fusion of time-of-flight depth and stereo for high accuracy depth maps. In: Proceedings of IEEE Conference on Computer Vision and Pattern Recognition (CVPR), pp. 1–8 (2008)

Printed by Printforce, the Netherlands